T0249464

Statistical Bioinformatics with R

Statistical Bioinformatics with R

Sunil K. Mathur
University of Mississippi

AMSTERDAM • BOSTON • HEIDELBERG • LONDON
NEW YORK • OXFORD • PARIS • SAN DIEGO
SAN FRANCISCO • SINGAPORE • SYDNEY • TOKYO

Academic Press is an imprint of Elsevier

Academic Press is an imprint of Elsevier
30 Corporate Drive, Suite 400, Burlington, MA 01803, USA
525 B Street, Suite 1900, San Diego, California 92101-4495, USA
84 Theobald's Road, London WC1X 8RR, UK

Notices
Knowledge and best practice in this field are constantly changing. As new research and experience broaden our understanding, changes in research methods, professional practices, or medical treatment may become necessary.

Practitioners and researchers must always rely on their own experience and knowledge in evaluating and using any information, methods, compounds, or experiments described herein. In using such information or methods they should be mindful of their own safety and the safety of others, including parties for whom they have a professional responsibility.

To the fullest extent of the law, neither the Publisher nor the authors, contributors, or editors, assume any liability for any injury and/or damage to persons or property as a matter of products liability, negligence or otherwise, or from any use or operation of any methods, products, instructions, or ideas contained in the material herein.

Library of Congress Cataloging-in-Publication Data
Mathur, Sunil K.
   Statistical bioinformatics with R / Sunil K. Mathur.
      p. cm.
   Includes bibliographical references and index.
   ISBN 978-0-12-375104-1 (alk. paper)
   1. Bioinformatics–Statistical methods. 2. R (Computer program language) I. Title.
   QH324.2.M378 2010
   570.285′5133–dc22

                                  2009050006

British Library Cataloguing-in-Publication Data
A catalogue record for this book is available from the British Library.

ISBN: 978-0-12-375104-1

For information on all Academic Press publications
visit our Web site at *www.elsevierdirect.com*

Printed and bound by CPI Group (UK) Ltd, Croydon, CR0 4YY

**Working together to grow
libraries in developing countries**

www.elsevier.com | www.bookaid.org | www.sabre.org

ELSEVIER   BOOK AID
               International   Sabre Foundation

Contents

Preface

Bioinformatics is an emerging field in which statistical and computational techniques are used extensively to analyze and interpret biological data obtained from high-throughput genomic technologies. Genomic technologies allow us to monitor thousands of biological processes going on inside living organisms in one snapshot, and are rapidly growing as driving forces of research, particularly in the genetics, biomedical, biotechnology, and pharmaceutical industries.

The success of genome technologies and related techniques, however, heavily depends on correct statistical analyses of genomic data. Through statistical analyses and the graphical displays of genomic data, genomic experiments allow biologists to assimilate and explore the data in a natural and intuitive manner. The storage, retrieval, interpretation, and integration of large volumes of data generated by genomic technologies demand increasing dependence on sophisticated computer and statistical inference techniques. New statistical tools have been developed to make inferences from the genomic data obtained through genomic studies in a more meaningful way.

This textbook is of an interdisciplinary nature, and material presented here can be covered in a one- or two-semester course. It is written to give a solid base in statistics while emphasizing applications in genomics. It is my sincere attempt to integrate different fields to understand the high-throughput biological data and describe various statistical techniques to analyze data. In this textbook, new methods based on Bayesian techniques, MCMC methods, likelihood functions, design of experiments, and nonparametric methods, along with traditional methods, are discussed. Insights into some useful software such as BAMarray, ORIGEN, and SAM are provided.

Chapter 1 provides some basic knowledge in biology. Microarrays are a very useful and powerful technique available now. Chapter 2 provides some knowledge of microarray technology and a description of current problems in using this technology. Foundations of probability and basic statistics, assuming that the reader is not familiar with statistical and probabilistic concepts, are provided in Chapter 3. Chapter 4 provides knowledge of some of the fundamental probability distributions and their importance in bioinformatics. Chapter 5 is about the inferential process used frequently in bioinformatics, and several tests and estimation procedures are described in that context. Some of the nonparametric methods such as the Wilcoxon and Mann-Whitney tests are useful in bioinformatics and are discussed in Chapter 6. Details of several well-known Bayesian procedures are provided in Chapter 7. Recently, Markov Chain Monte Carlo methods have been used effectively in bioinformatics, and Chapter 8 gives details of MCMC methods in a bioinformatics context. Chapter 9 is about analysis of variance methods that provide an important framework for the analysis of variance. Chapter 10 is about various statistical designs, including modern and classical designs found useful in bioinformatics. Chapter 11 discusses multiple comparison methods and error rates in a bioinformatics context. R-code for examples provided in the text is shown along with respective examples in each chapter. Real data are used wherever possible. Although this textbook is not a comprehensive account of techniques used in bioinformatics, it is my humble attempt to provide some knowledge to the audience in the area of bioinformatics.

AIMS AND OBJECTIVES OF THIS TEXTBOOK

This textbook covers theoretical concepts of statistics and provides a sound framework for the application of statistical methods in genomics. It assumes an elementary level of statistical knowledge and a two-semester sequence of calculus. Most of the textbooks at this level assume an advanced level statistical course and advanced level calculus course, which creates problems for nonstatistics majors and graduate students from nonstatistics disciplines who would like to take a course in statistical bioinformatics to gain knowledge in the interdisciplinary field. This textbook tries to provide a remedy for that problem. Statistics itself is an interdisciplinary applied subject; therefore, many real-life data examples and data are provided to explain the statistical methods in the genomics field. Codes for computer language R are provided for almost every example used in the textbook. Due to the tremendous amount of data, knowledge of using computers for data analysis is highly desirable.

One of the main aims of this textbook is to introduce concepts of statistical bioinformatics to advanced undergraduate students and beginning graduate students in the area of statistics, mathematics, biology, computer science, and other related areas, with emphasis on interdisciplinary applications and use of the programming language R in bioinformatics. This textbook is targeted to senior regular full-time undergraduate level and beginning graduate level students from statistics, mathematics, engineering, biology, computer science, operation research, pharmacy, and so forth. This textbook also is targeted to researchers who would like to understand different methods useful in understanding and analyzing high-throughput data. This textbook is also useful for people in the industry as a reference book on statistical bioinformatics, providing a solid theoretical base in statistics and a clear approach of how to apply statistical methods for analyzing genomic data, and the use of the computer programming language R in the implementation of the proposed procedures.

The prerequisite for this course is a semester of statistics at the elementary level; however, a higher-level course in statistics is desirable. The material in this textbook can be covered in two semesters. In this textbook, an attempt is made to provide theoretical knowledge to students and develop students' expertise to use their knowledge in real-world situations. At the same time, this textbook provides knowledge for using computer-programming skills in R to implement the methodology provided here.

In this textbook, concepts are developed from ground zero level to the advanced level. If desired, some chapters can be skipped from a teaching point of view, and the remaining material can be covered in one semester. The following flowchart provides options on how to use this textbook for teaching a one-semester course. For a two-semester course, coverage of the complete textbook is recommended.

This textbook is the outcome of classroom teaching of the senior undergraduate and beginning graduate level course in statistical bioinformatics for several semesters, seminars, workshops, and research in the area of statistical bioinformatics. Material in this textbook has been tested in classroom teaching of a one-semester course for several years. The audience was senior undergraduate students and beginning graduate students from many disciplines who had two semesters of calculus, and most of them had an elementary level statistical background.

Codes for R programming provided in this textbook are available on the publisher's website under the link for this textbook. R programs provided in the textbook are basic in nature, and are included just to give an idea about the use of R. Users can improve these programs and make them more elegant and sophisticated. Exercises and solution manuals are available on the publisher's website.

KEY FEATURES

- Bioinformatics is an interdisciplinary field; therefore, the audience is from different fields and different backgrounds. It is hard to assume that the audience will have a good background in statistics or biology. Therefore, this textbook presents the concepts from ground zero level and builds it to an advanced level.

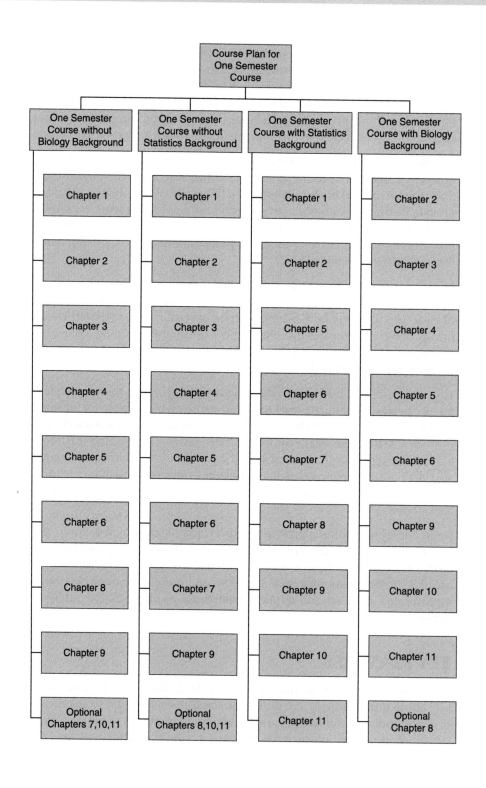

- Methods presented in the textbook are based on well-documented research papers. References are provided in the text, and an extensive bibliography is presented at the end of the textbook.

- Depending on the background of the audience, material can be covered in a one-semester course in various ways as presented in the previous flowchart.

- Statistical concepts are presented to give a solid foundation in statistics, and methods are illustrated through the use of several relevant examples.

- Exercises provided on the publisher's website are a good mix of theory and real-life applications.

- The applications of the methods in this textbook are shown using the programming language R. R, which is a free language, is widely used in statistics. This will help students to utilize their computer programming skills in applying statistical methods to study real data in a bioinformatics context.

- Modern computational methods presented in the textbook such as the Markov Chain Monte Carlo method have been found to be very useful in analyzing genomic data.

- A chapter is devoted to the useful and popular concept of Bayesian statistics with applications.

- A chapter is provided on the design of experiments, including several of the new designs with applications.

- A chapter is devoted to multiple comparisons.

- For an instructor's solution manual and additional resources, please visit www.elsevierdirect.com and http://textbooks.elsevier.com.

Acknowledgments

This textbook would not have been possible without the active support of many people known to me and many unknown. I am indebted to my friend and mentor Dr. Shyamal Peddada, who continuously guided and mentored me throughout my academic life. I am grateful to my Ph.D. supervisor, Dr. Kanwar Sen of University of Delhi, who inculcated my interest in statistics; otherwise, I would have gone to industry for sure. I am grateful to my iconic guides and philosophers Dr. C.R. Rao and Dr. P.K. Sen for their useful discussions with me. They were there whenever I needed them. I am thankful to Dr. Sunil J. Rao and Dr. Ajit Sadana for going through parts of the manuscript even when they were busy with their own important projects. I thank my friends, collaborators, and colleagues, including Dr. Dawn Wilkins, University of Mississippi; Dr. Ed Perkins, U.S. Corp of Engineers; and Dr. E.O. George, University of Memphis. My thanks go to a number of reviewers who took time to go through the manuscript and make suggestions, which led to improvement in the presentation of this textbook. I thank my students who took a course in statistical bioinformatics under me and were made guinea pigs for testing the material presented in this textbook. I thank the R-core team and numerous contributors to R. Several of the figures used in the textbook are courtesy of the National Institutes of Health.

My sincere thanks to my former chancellor, Dr. Robert Khayat of the University of Mississippi, whose leadership qualities always motivated me. My sincere thanks go to Dr. Glenn Hopkins, dean of my school, and Dr. Bill Staton, for all their support and encouragement.

This textbook would not have been possible without the support and encouragement of my beautiful, intelligent, and loving wife, Manisha. She provided so much in the last few years that I may not be able to fully make it up to her for rest of my life. This work would not have been feasible without the cooperation of my kids, Shreya, and Krish, who sacrificed many personal hours that enabled me to work on this textbook. I am thankful to them. I am also thankful to my family, especially my parents, Mr. O. P. Mathur and

Mrs. H. Mathur, and in-laws, Mr. R.P. Mathur and Mrs. S. Mathur, for their continuous encouragement.

This textbook would not have taken its shape in print form without the support of Lauren Schultz, senior acquisition editor at Elsevier. She was very cooperative and encouraging. It was great to work with her, and I am grateful to her. My thanks also go to Mr. Gavin Becker, assistant editor, who took responsibility to see this textbook through the production process. In addition, I am thankful to Christie Jozwiak, project manager for this textbook, and the other staff at Elsevier who took personal care of this textbook.

Many people helped me to complete this task. I am deeply indebted to those people without whom I would not have been able to even start this textbook.

Sunil K. Mathur

## CHAPTER 1

Introduction

CONTENTS

1.1 STATISTICAL BIOINFORMATICS

Recent developments in high-throughput technologies have led to explosive growth in the understanding of an organism's DNA, functions, and interactions of biomolecules such as proteins. With the new technologies, there is also an increase in the databases that require new tools to store, analyze, and interpret the information contained in the databases. These databases contain complex, noisy, and probabilistic information that needs adaptation of classical and nonclassical statistical methods to new conditions. The new science of statistical bioinformatics (Figure 1.1.1) aims to modify available classical and nonclassical statistical methods, develop new methodologies, and analyze the databases to improve the understanding of the complex biological phenomena. This interdisciplinary science requires understanding of mathematical

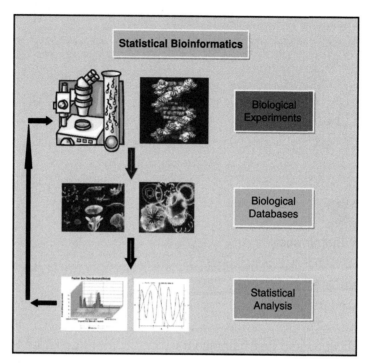

FIGURE 1.1.1

Statistical bioinformatics. (See color insert).

and statistical knowledge in the biological sciences and its applications. The research in statistical bioinformatics is advancing in all directions from the development of new methods, modification of available methods, combination of several interdisciplinary tools, to applications of statistical methods to high-throughput data. When one looks at the growth in the number of research papers in statistical bioinformatics published in leading statistical journals, it appears that statistical bioinformatics has become the most prominent field in statistics, where every discipline in statistics has contributed to its advancement. Statistical bioinformatics is so fascinating that the number of research papers published in this area is increasing exponentially.

There are a number of fields that are getting benefits from the developments in statistical bioinformatics. It helps to identify new genes and provide knowledge about their functioning under different conditions. It helps researchers to learn more about different diseases such as cancer and heart disease. It provides information about the patterns of the gene activities and hence helps to further classify diseases based on genetic profiles instead of classifying them based on the organs in which a tumor or disease cells are found. This enables the pharmaceutical companies to target directly on the gene specific or tumor specific to find the cure for the disease. Statistical bioinformatics helps to understand the correlations between therapeutic responses to drugs and genetic profiles, which helps to develop drugs that counter the proteins produced by specific genes and reduce or eliminate their effects. Statistical bioinformatics helps to study the correlation between toxicants and the changes in the genetic profiles of the cells exposed to such toxicants.

1.2 GENETICS

Genetics occupies a central role in biology. Any subject connected with human health requires an understanding of genetics. It is necessary to have sufficient knowledge of genetics to understand the information contained in the biological data.

Genetics is the study of heredity. Heredity is the process by which characteristics are passed from one generation to the next generation. The concept of heredity was observed by people long before the principles and procedures of genetics were studied. People wondered how a son looks similar to his father, why two siblings look different, and why the same diseases run across generations. In 1860, Mendel performed the first experimental work in the field to establish that traits are transmitted from one generation to another systematically. He studied the inheritance of seven different pairs of contrasting characters in a garden pea (*Pisum sativum*) plant by taking one pair at a time. He crossed two pea plants with alternate characters by artificial pollination, which led to the identification of two principles of heredity. Let T denote a tall parent and d denote a dwarf parent. The parental plants are denoted by P_1. The progeny obtained after crossing between parents is called first filial (offspring) generation and is denoted by F_1. The second filial generation obtained as a result of self-fertilization among F_1 plants is represented by F_2. The monohybrid cross is represented by the Figure 1.2.1.

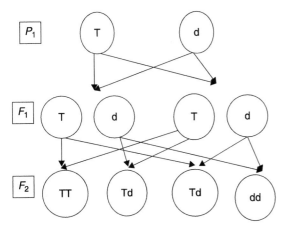

FIGURE 1.2.1
Monohybrid cross.

The tall and dwarf plants in F_2 are in the approximate ratio of 3:1. The dwarf plants in the F_2 filial produced only dwarf plants in the next generation. The tall plants of the F_2 generation produced only one-third tall plants in next generation. The other two-thirds produced tall and dwarf plants in the ratio of 3:1.

Thus, we have the following proportions in the F_2 generation:

Tall homozygous	25% (TT)
Tall heterozygous	50% (Td)
Dwarf homozygous	25% (dd).

Based on experimental results, Mendel postulated three laws, which are known as Mendel's three laws of heredity:

1. Law of dominance and recessiveness
2. Law of segregation
3. Law of independent assortment.

It is observed that genetic factors exist in pairs in individual organisms. In the peas experiment, there would be three possible pairwise combinations of the two factors.

When two unlike factors for a single characteristic are present, one factor is *dominant* as compared to the other. The other factor is considered *recessive*. In the peas experiment, the factor for tall is dominant to the factor for dwarf.

During gamete formation, factors segregate randomly so that each gamete receives one or the other with equal likelihood. The offspring of an individual with one of each type of factor has an equal chance of inheriting either factor.

The term *genetics* was first used by Bateson in 1906, which was derived from the Greek word *gene* meaning "to become" or "to grow into." Thus, genetics is the science of coming into being. It studies the transmission of characteristics from generation to generation. Mendel assumed that a character is determined by a pair of factors or determiners that are present in the organism. These factors are known as genes in modern genetics.

The term *genotype* denotes the genetic makeup of an organism, whereas the term *phenotype* is the physical appearance of an individual. In Mendel's peas experiment, the phenotype ratio of F_2 is 3:1, while the genotype ratio of F_2 is 1:2:1.

If, in an organism, two genes for a particular character are identical, then the organism is known as homozygous. If the two genes are contrasting, then it is known as heterozygous. An allele is a mutated version of a gene. These represent alternatives of a character. A heterozygote possesses two contrasting genes or alleles, but only one of the two is able to express itself, while the other will remain unexpressed. The expressed gene is known as a dominant gene, while the unexpressed gene or allele is known as a recessive gene. A monohybrid cross involves the study of inheritances of one pair of contrasting characters. The study of characteristics of tall and dwarf peas by Mendel is a classic example of monohybrid cross. A test cross is used to determine the genotype of an individual exhibiting a dominant phenotype. The test score parent is always homozygous recessive for all of the genes under study. The test cross will determine how many different kinds of gametes are being produced.

■ Example 1

Let us consider the coat colors in a guinea pig. Let us cross a homozygous black guinea pig with a homozygous albino guinea pig (Figure 1.2.2).

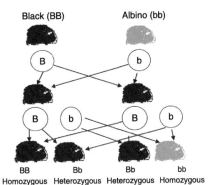

FIGURE 1.2.2

Crossing of a homozygous black guinea pig with a homozygous albino guinea pig.

Thus, in F_2 generation, black and white guinea pigs are in the ratio of 3:1.

Thus, dominant and recessive characters can be found in various organisms. For example, in tomatoes, a purple stem color is a dominant character, while the green stem color is a recessive color. The tall plant is dominant, while the dwarf plant is recessive. ■

■ Example 2

Let us assume that gene G is dominant over gene g.

Find the phenotypic ratio in first-generation offspring obtained after crossing

    (a) Gg X gg

    (b) GG X gg.

Solution

This is a case of monohybrid cross. Let us solve the problem by assuming this is a checkerboard problem.

(a)

Gg X gg

	g	g
G	Gg	Gg
g	gg	gg

Now we observe from the table that Gg and gg are in the ratio of 1:1.

(b)

GG X gg

	g	g
G	Gg	Gg
G	Gg	Gg

Thus, Gg is 100% in the first offspring.   ■

1.3 CHI-SQUARE TEST

The Chi-square goodness of fit test is used to find whether the observed frequency obtained from an experiment is close enough to expected frequency.

If $O_i, i = 1, 2, \ldots, n$ denotes the observed (experimental) frequency of the subjects, and if $E_i, i = 1, 2, \ldots, n$ is the expected frequency of the subjects, n is the sample size, then the Chi-square test can be defined as follows:

$$\chi^2 = \sum_{i=1}^{n} \frac{(O_i - E_i)^2}{E_i}. \tag{1.3.1}$$

The rejection criteria are that if the observed value of Chi-square is greater than the critical value of Chi-square, we reject the claim that the observed frequency is the same as that of the expected frequency (for more details, see Chapter 5).

■ Example 1

In an experiment a scientist crossed a white flower plant with a green flower plant. In the F_2 generation, the scientist gets 75 plants with green flowers and 29 plants with white flowers. The ratio of green to white in this cross is 2.68:1. Mendel's ratio was 3:1. The scientist is wondering whether this is the same ratio, close enough to match Mendel's ratio.

Solution

We calculate the expected number of green flowers and white flowers.

Expected number of green flowers

$= E \text{ (green flowers)}$

$= $ Ratio of green flowers multiplied by number of total flowers in the

F_2 generation

$= (3/4) \cdot 104 = 78.$

Similarly,

$$E \text{ (white flowers)} = (1/4) \cdot 104 = 26.$$

Since there are two types of flowers involved, $n = 2$.

Also, we find $O_1 = 75$ and $O_2 = 29$ while $E_1 = 78$ and $E_2 = 26$.

Thus,

$$\chi^2 = \sum_{i=1}^{2} \frac{(O_i - E_i)^2}{E_i} = \frac{(75 - 78)^2}{78} + \frac{(29 - 26)^2}{29}$$

$$= 0.4257.$$

Now we look up a table of Chi-square critical values, at the degrees of freedom $n - 1$, which is $2 - 1 = 1$ because $n = 2$, and find that the critical value $\chi^2_{\text{(critical)}}$ is 3.841.

Here, we have $\chi^2_{\text{(Observed)}} = 0.4257 < \chi^2_{\text{(critical)}} = 3.841$; therefore, the scientist's claim that the observed ratio is close enough to the expected ratio is accepted.

The R-code for example 1.3.1 is as follows:

```
> Example 1.3.1
> green<-75
> white<-29
> sum<-green+white
> ratio-of-green<-3/4
> ratiogreen<-3/4
> ratiowhite<-1/4
> expectedgreen<-ratiogreen*sum
> cat("Expected number of Green Flowers:",expectedgreen,fill = T)
Expected number of Green Flowers: 78
> expectedwhite<-ratiowhite*sum
> cat("Expected number of Green Flowers:",expectedwhite,fill = T)
Expected number of Green Flowers: 26
> cat("Expected number of White Flowers:",expectedwhite,fill = T)
```

Expected number of White Flowers: 26
```
> 01<-75
> 02<-29
> e1 = expectedgreen
> e2 = expectedwhite
> chi<-((01-e1)^2/e1)+((02-e2)^2/e2)
> cat("Calculated value of Chi-square from data:",chi,fill = T)
Calculated value of Chi-square from data: 0.4615385
```

It can also be done in the following way.

```
> chisq.test(c(75,29),p = c(3/4,1/4))
      Chi-squared test for given probabilities
data: c(75, 29)
X-squared = 0.4615, df = 1, p-value = 0.4969
```

■

■ Example 2

A girl has a rare abnormality of the eyelids called ptosis. The ptosis makes it impossible for the eyelid to open completely. The gene P is found to be dominant in this condition. The girl's father had ptosis and the father's mother had normal eyelids, but the girl's mother did not have this abnormality.

(a) If the girl marries a man with normal eyelids, what proportion of her children will have this abnormality?

(b) Say the girl has four offspring who have normal eyelids while five offspring have ptosis in the F_2 generation. Does this outcome confirm the theoretical ratio between ptosis and normal offspring in the F_2 generation?

Solution

(a) Let the recessive gene be p. Then the genotype can be displayed (Figure 1.3.1) as follows:

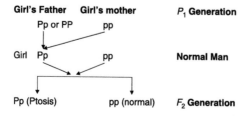

FIGURE 1.3.1

Genotype of ptosis.

Thus, 50% of her offspring will be with ptosis, while 50% will be normal.

(b) According to the genotype, the expected ratio of offspring with ptosis to normal offspring is 1:1.
Therefore, we have

$$O_1 = 4 \text{ and } O_2 = 5, \text{ while } E_1 = 4.5 \text{ and } E_2 = 4.5.$$

Thus,

$$\chi^2 = \sum_{i=1}^{2} \frac{(O_i - E_i)^2}{E_i} = 0.444.$$

Here, we have $\chi^2_{(Observed)} = 0.444 < \chi^2_{(critical)} = 3.841$; therefore, we accept the claim that the observed frequency is close enough to the expected frequency. Thus, the theoretical ratio between ptosis and normal offspring in the F_2 generation is the same.

The R-code is as follows:

```
> chisq.test(c(4,5),p = c(1/2,1/2))
    Chi-squared test for given probabilities
data: c(4, 5)
X-squared = 0.1111, df = 1, p-value = 0.7389
Warning message:
In chisq.test(c(4, 5), p = c(1/2, 1/2)):
Chi-squared approximation may be incorrect
```

∎

1.4 THE CELL AND ITS FUNCTION

The cell is the smallest entity of life. Organisms contain multiple units of cells to perform different functions. Some organisms, such as bacteria, may have only a single cell to perform all their operations. But other organisms need more than just a single cell. The cell is enclosed by a plasma membrane, inside which all the components are located to provide the necessary support for the life of the cell.

There are two basic types of cells, which are known as prokaryotic cells and eukaryotic cells. The prokaryotic cells have no nucleus. These types of cells are generally found in bacteria. The eukaryotic cells have true nuclei. Plants and animals have eukaryotic cells. Most eukaryotic cells are diploid, which means that they contain two copies, or homologues, of each chromosome. The gametes, which are reproductive cells and are haploid, have only one copy of each chromosome. The eukaryotic cells also have organelles, each of which is

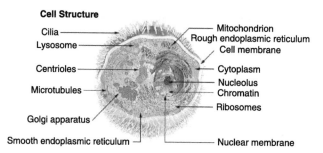

Cell Structure

Cilia
Lysosome
Centrioles
Microtubules
Golgi apparatus
Smooth endoplasmic reticulum

Mitochondrion
Rough endoplasmic reticulum
Cell membrane
Cytoplasm
Nucleolus
Chromatin
Ribosomes
Nuclear membrane

FIGURE 1.4.1

Cells (Courtesy: National Cancer Institute-National Institutes of Health). (See color insert).

specialized to perform a specific function within the cell. The genetic material of cells is organized into discrete bodies called chromosomes. The chromosomes are a complex of DNA and protein.

Most eukaryotic cells divide mitotically, progressing regularly through the cell cycle. The cell cycle consists of a series of stages that prepare the cell for division (primarily by replicating the chromosomes). Meiosis is the process by which cells divide to produce gametes, such as pollens and ovums in plants, sperms and eggs in animals. The main purpose of meiosis is to reduce the number of chromosomes in the cells by half.

The cells of most eukaryotes are diploid; that is, they have two of each type of chromosome. Meiosis produces cells that are haploid, and haploids, as mentioned earlier, have only one of each type of chromosome.

Meiosis is divided into stages, much like mitosis. Unlike mitosis, however, meiosis is a two-step process, consisting of two sequential cell divisions. Therefore, one basic difference between meiosis and mitosis is that at the end of meiosis, there are four haploid progeny cells, whereas at the end of mitosis, there are two diploid progeny cells. Two cell divisions in meiosis are called meiosis I and meiosis II. Each of these divisions is divided into the same stages as mitosis—namely, prophase, metaphase, anaphase, and telophase. Stages are numbered according to which meiotic cell division is being discussed. For example, prophase of the first meiotic cell division is called prophase I; anaphase of the second meiotic cell division is called anaphase II.

The different organelles within the cell, as shown in Figure 1.4.1, are described next.

Endoplasmic Reticulum (ER)

The endoplasmic reticulum, or ER, comes in two forms: smooth and rough. The surface of the rough ER is coated with ribosomes. The functions of ER include mechanical support, synthesis (especially proteins by rough ER), and transport.

Golgi Complex

The Golgi complex consists of a series of flattened sacs (or cisternae). Its functions include synthesis (of substances likes phospholipids), packaging of materials for transport (in vesicles), and production of lysosomes.

Lysosomes

Lysosomes are membrane-enclosed spheres that contain powerful digestive enzymes. Their functions include destruction of damaged cells, which is why they are sometimes called "suicide bags." The other important function is the digestion of phagocytosed materials such as mitochondria. Mitochondria have a double-membrane: an outer membrane and a highly convoluted inner membrane.

Inner Membrane

The inner membrane of mitochondria has folds or shelf-like structures called cristae that contain elementary particles. These particles contain enzymes important in production of adenosine triphosphate (ATP). Its primary function is ATP.

Ribosomes

Ribosomes are composed of ribosomal ribonucleic acid (rRNA). Their primary function is to produce proteins. Proteins may be dispersed randomly throughout the cytoplasm or attached to the surface of rough endoplasmic reticulum often linked together in chains called polyribosomes or polysomes.

Centrioles

Centrioles are paired cylindrical structures located near the nucleus. They play an important role in cell division.

Flagella and Cilia

The flagella and cilia are hair-like projections from some human cells. Cilia are relatively short and numerous (e.g., those lining trachea). A flagellum is relatively long, and there is typically just one (e.g., sperm) in a cell.

Nucleus (Plural Nuclei)

Cells contain a nucleus, which contains deoxyribonucleic acid (DNA) in the form of chromosomes, plus nucleoli within which ribosomes are formed.

The nucleus is often the most prominent structure within a eukaryotic cell. It maintains its shape with the help of a protein skeleton known as the nuclear matrix. The nucleus is the control center of the cell. Most cells have a single nucleus, but some cells have more than one nucleus. The nucleus is surrounded by a double layer membrane called the nuclear envelope. The nuclear envelope is covered with many small pores through which protein and chemical messages from the nucleus can pass. The nucleus contains DNA, the hereditary material of cells. The DNA is in the form of a long strand called chromatin. During cell division, chromatin strands coil and condense into thick structures called chromosomes. The chromosomes in the nucleus contain coded information that controls all cellular activity. Most nuclei contain at least one nucleolus (plural *nucleoli*). The nucleolus makes (synthesizes) ribosomes, which in turn build proteins. When a cell prepares to reproduce, the nucleolus disappears.

1.5 DNA

The genome is a complete set of instructions to create and sustain an organism. The DNA is the support for the genome. Genes are specific sequences of nucleotides that encode the information required for making proteins (Figure 1.5.3). In 1868, Johann Miesher found an acidic substance isolated from human pus cells, which he called "nuclein." In the 1940s, the existence of polynucleotide chains was documented, and the role of nucleic acid in storing and transmitting the genetic information was established in 1944. There are two types of nucleic acid: deoxyribonucleic acid (DNA) and ribonucleic acid (RNA). In most organisms, the DNA structure carries the genetic information, whereas in some organisms, such as viruses, RNA carries the genetic information.

There are four different bases found in DNA—namely, adenine (A), guanine (G), thymine (T), and cytosine (C). RNA has adenine, guanine, and cytosine but has uracil (U), in place of thymine. Purines include adenine and guanine, which are double-ring based, while pyrimidines include cytosine, thymine, and uracil. RNA has a single-stranded polymer structure (Figure 1.5.1), but DNA has a double-stranded structure.

At Cambridge University, a British student, Francis Crick, and an American visiting research fellow, James Watson, studied the structure of DNA. At the same time, at King's College, London, Rosalind Franklin and Maurice Wilkins studied DNA structure. Franklin was able to use the X-ray to find a picture of the molecule's structure. Wilkins shared the X-ray pictures with Watson without the knowledge of Franklin (Sayre, 2000; Maddox, 2002). Based on this information, Watson and Creek published a paper in *Nature* proposing that the DNA molecule was made up of two chains of nucleotides paired in such a way as to form a double helix, like a spiral staircase (Watson and Crick, 1953). Watson, Crick, and Wilkins received the Nobel Prize in 1962. Despite Franklin's important contribution to the discovery of DNA's helical structure, she could not be named a prizewinner because four years earlier she had died of cancer at the age of 37, and the Nobel Prize cannot be given posthumously.

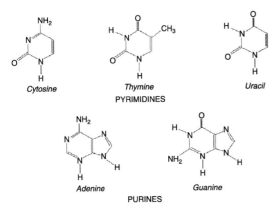

FIGURE 1.5.1

Pyrimidines and purines.

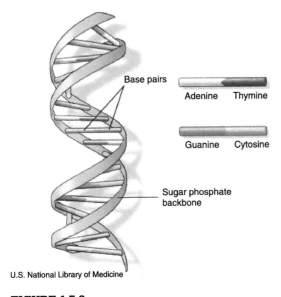

U.S. National Library of Medicine

FIGURE 1.5.2

DNA. (See color insert).

In the DNA structure, each polynucleotide chain consists of a sequence of nucleotides linked together by phosphate bonds, which join deoxyribose moieties (Figure 1.5.2). The hydrogen bonds hold two polynucleotide strands in their helical configuration. Bases are between two nucleotide chains like steps. The base pairing is done as follows: Adenine is paired with thymine, and guanine is paired with cytosine. All base pairs consist of one purine and one pyrimidine. The location of each base pair depends on the hydrogen

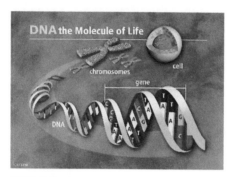

FIGURE 1.5.3

DNA, the molecule of life (http://genomics.energy.gov). (See color insert).

bonding capacity of the base. Adenine and thymine form two hydrogen bonds, and guanine and cytosine form three hydrogen bonds. Cytosine and adenine or thymine and guanine do not form hydrogen bonding in a normal configuration. Thus, if the sequence of bases in one strand is known, the sequence of bases in other strands can be determined. The phosphate bonds in one strand go from 3′ carbon of one nucleotide to 5′ carbon of the adjacent nucleotide, while in the complimentary strand the phosphodiester bonds go from 5′ carbon to 3′ carbon.

As mentioned earlier, RNA is single-stranded, whereas DNA is double-stranded. RNA contains the pyrimidine uracil (U), and it pairs with adenine (A). RNA plays an important role in protein synthesis and acts as a messenger, known as mRNA, carrying coded information from DNA to ribosomal sites of protein synthesis in cells. Ribosomal RNA (rRNA) contains the bulk of RNA, and transfer RNA (tRNA) attaches to amino acids.

1.6 DNA REPLICATION AND REARRANGEMENTS

DNA is replicated when a cell divides itself. If the replication is not exact, it causes mutation of genes. During the replication, the hydrogen bonds between the bases break, which leads nucleotide chains of DNA to separate and uncoil. Each nucleotide of separated chain attracts its complement from the cell and forms a new nucleotide chain. The double helix structure of the DNA serves as a template for the replication. Each daughter DNA molecule is an exact copy of its parent molecule, which consists of one parental strand and one new DNA strand (Figure 1.6.1). This type of the replication is called semi-conservative replication. About 20 or more different proteins and enzymes are required during the replication, which is known as the replisome or DNA replicase system.

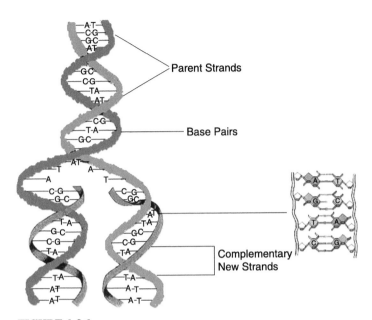

FIGURE 1.6.1

DNA replication. (See color insert).

1.7 TRANSCRIPTION AND TRANSLATION

The genetic information contained in the DNA is transcribed into messenger RNA (mRNA). DNA unfolds, and one of the two strands acts as a template for the synthesis of a sequence of nucleotides. The synthesis of mRNA proceeds in the 5′ to 3′ direction. Transcription occurs in three basic steps: initiation, elongation, and termination. The recognition of the promoter sequence starts with the initiation in prokaryotes by the sigma subunit of RNA polymerase while in the case of eukaryotes it is started by the transcription factors, which then recruit RNA polymerase to the promoter. In prokaryotes, termination occurs at termination sequences, which depend on the gene that may or may not involve the termination protein rho. In eukaryotes, termination is less specific, and it occurs well downstream of the 3′ end of the mature mRNA.

DNA → Transcription → mRNA → Translation → Proteins.

The synthesis of amino acids by using the information encoded in the mRNA molecule is called translation. This process involves mRNA, tRNA, and ribosomes. Ribosomes are large organelles made of two subunits, each of which is composed of ribosomal RNA and proteins. The tRNA (transfer RNA) contains anticodons and has a specific location on mRNA. Inside the cytoplasm, it gathers the amino acid, specified by the codon, and it attaches to mRNA at a specific codon. The unique structure of tRNA exposes an anticodon that binds to codons in an mRNA and an opposite end that binds to a specific amino acid. The aminoacyl tRNA enzyme binds an amino acid to a tRNA in a process called charging. The translation happens in three steps: initiation, elongation, and termination. In initiation, the formation of the ribosome/mRNA/initiator tRNA complex is achieved, while in elongation actual synthesis of the polypeptide chain is done by the formation of peptide bonds between amino acids. In the last step, termination dissociates the translation complex and releases the finished polypeptide chain.

1.8 GENETIC CODE

It was known that genes pass genetic information from one generation to another, but the real question is how the genes store this genetic information. Experiments done by Avery and Macleod, and McCarty, Hershey, and Chase show that the genetic information is contained in genes. The Watson and Crick Model does not offer any obvious explanation regarding how the genetic information is stored in the DNA. Actually, the genetic information is contained in the genetic code, which contains the directions on how to make proteins and what sequence of proteins to be made. This sequence of proteins determines the properties of cells and hence overall characteristics of an organism.

A codon is the nucleotide or nucleotide sequence in the mRNA that codes for a particular amino acid, whereas the genetic code is the sequence of nitrogenous bases in an mRNA molecule, which contains the information for the synthesis of the protein molecule. We know that the nucleotides A, T, G, and C make up DNA, but we do not know how the sequence of these four letters determines the overall characteristics of an organism. It was found that there are 20 different amino acids in proteins, and a combination of only one or two nitrogenous bases cannot provide sufficient code words for 20 amino acids, but if we keep a codon of three nucleotides out of four nucleotides, we will get $4^3 = 64$ possible combinations that will easily cover 20 amino acids. This type of argument led to the triplet code of genetics. Thus, a sequence of three nucleotides (codons) in mRNA makes an amino acid in proteins. The triplets are read in order, and each triplet corresponds to a specific protein (Figure 1.8.1). After one triplet is read, the "reading frame" shifts over three letters, not just one or two letters. Through use of a process called in vitro synthesis, the codons for all 20 amino acids have been identified.

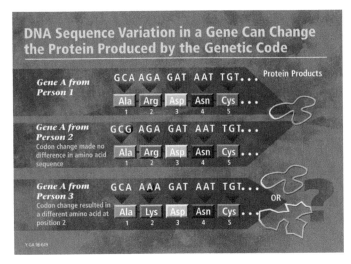

FIGURE 1.8.1

DNA sequence (http://genomics.energy.gov). (See color insert).

■ Example 1

How many triplet codons can be formed from four ribonucleotides A, U, G, and C containing (a) no cytosine? (b) one or more cytosine?

Solution

(a) The cytosine is one of four nucleotides; thus, the probability that the cytosine will be in the codon is $1/4$, while the probability that it will not be a letter in the codon is $3/4$.
Thus,

$$P \text{ (none of the three letters in a codon is a cytosine)}$$

$$= \frac{3}{4} \cdot \frac{3}{4} \cdot \frac{3}{4} = \left(\frac{3}{4}\right)^3 = \frac{27}{64}.$$

Thus, 27 triplet codons can be formed.

(b) P codon has at least one cytosine

$$= 1 - P \text{ (codon does not have any cytosine)}$$

$$= 1 - \frac{27}{64} = \frac{37}{64}.$$

Thus, 37 codons can be formed with one or more cytosines.

The R-code for Example 1 is as follows:

```
> p<-(3/4)^3
> cat("P(None of three letters in a codon is a cytosine) = ",p,fill = T)
P(None of three letters in a codon is a cytosine) = 0.421875
> n1<-(3/4)^3*64
> cat("Number of triplet codes which can be formed not having a cytosine) = ",n1,fill = T)
Number of triplet codes which can be formed not having a cytosine) = 27
p<-(3/4)^3
> q<-1-p
> n2<-q*64
> cat("P(Codons having at least one cytosine = ",q,fill = T)
P(Codons having at least one cytosine = 0.578125
> cat("Codons having at least one cytosine = ",n2,fill = T)
Codons having at least one cytosine = 37
```

■

The genetic code is triplet in nature. The AUG is a start codon, which sets the reading frame and signals the start of the translation of the genetic code. After AUG, the translation continues in a nonoverlapping fashion until a stop codon (UAA, UAG, or UGA) is encountered in a frame. The nucleotides between the start and stop codons comprise an open-reading frame. Thus, the whole genetic code is read and the corresponding amino acid is identified. Each codon is immediately followed by the next codon without any separation. The genetic codes are nonoverlapping, and a particular codon always codes for the same amino acid. The same amino acid may be coded by more than one codon, but a codon never codes for two different amino acids.

■ Example 2

Using the information in Table 1.8.1, convert the following mRNA segment into a polypeptide chain. Does the number of codons equal the number of amino acids?

5' AUG GTT GCA UCA CCC UAG 3'

Solution

met – val – ala – ser – pro – (terminate)

The number of codons is six, and the number of amino acids is five because the last codon UAG is a terminating codon.　■

Table 1.8.1 Triplet Codons for Amino Acid

F		Second Position					T
I		**T**	**C**	**A**	**G**		H
R	T	TTT Phe	TCT Ser	TAT Tyr	TGT Cys	T	I
S		TTC Phe	TCC Ser	TAC Tyr	TGC Cys	C	R
T		TTA Leu	TCA Ser	TAA Ter	TGA Ter	A	D
		TTG Leu	TCG Ser	TAG Ter	TGG Trp	G	
P	C	CTT Leu	CCT Pro	CAT His	CGT Arg	T	P
O		CTC Leu	CCC Pro	CAC His	CGC Arg	C	O
S		CTA Leu	CCA Pro	CAA Gln	CGA Arg	A	S
I		CTG Leu	CCG Pro	CAG Gln	CGG Arg	G	I
T	A	ATT Ile	ACT Thr	AAT Asn	AGT Ser	T	T
I		ATC Ile	ACC Thr	AAC Asn	AGC Ser	C	I
O		ATA Ile	ACA Thr	AAA Lys	AGA Arg	A	O
N		ATG Met	ACG Thr	AAG Lys	AGG Arg	G	N
	G	GTT Val	GCT Ala	GAT Asp	GGT Gly	T	
		GTC Val	GCC Ala	GAC Asp	GGC Gly	C	
		GTA Val	GCA Ala	GAA Glu	GGA Gly	A	
		GTG Val	GCG Ala	GAG Glu	GGG Gly	G	

1.9 PROTEIN SYNTHESIS

Proteins are the most important molecules in the biological system because they not only form the building material of the cell, but are also responsible for controlling all the biochemical reactions present in the living system as enzymes. Proteins are also responsible for controlling the expression of phenotypic traits.

There are 20 amino acids, but there are several thousands of proteins, which are formed by various arrangements of 20 amino acids. In a sequence of more than 100 amino acids, even if a single amino acid is changed, the whole function of resulting proteins may change or even destroy the cell. Protein synthesis is nothing but the arrangement of a definite number of amino acids in a particular sequential order. The sequence of amino acids is determined by the sequence of nucleotides in the polynucleotide chain of DNA. As mentioned earlier, a sequence of three nucleotides is called codon or triplet code.

The flow of genetic information (Figure 1.9.1) for the protein synthesis was proposed by Crick (1958).

In the genetic information flow, the transcription converts DNA to mRNA, and then through translation, mRNA produces proteins. This is known as a

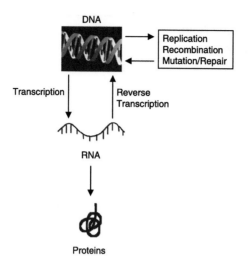

FIGURE 1.9.1

Genetic flow of information. (See color insert).

one-way flow of information or central dogma. During the transcription of mRNA from DNA, the DNA-dependent RNA polymerase enzyme, the genetic message encoded in DNA, is transcribed into mRNA. The two strands of the specific DNA unfolds, and one of these two strands becomes a template (the daughter strand) from which the exact sequence of nucleotides is transmitted to the mRNA molecule. Through translation, the sequence of nucleotides in mRNA is translated into the sequence of amino acids of a polypeptide chain. The termination of a polypeptide chain occurs when the ribosome carrying peptidyl tRNA reaches any of the termination codons like UAA, UAG, or UGA that are present on mRNA at the end of the cistron.

EXERCISE 1

For questions 1 through 15, fill in the blanks.

1. How many strands make up a double helix?
2. _____ holds one strand against the other in the double helix.
3. _____ is a strand of DNA.
4. _____ are the four bases of DNA that form in the double helix.
5. Which bases are purines?
6. Which bases are pyrimidines?
7. Where is DNA located in the cell?
8. _____ is used in the production of ATP.

9. _____ is a basic unit of heredity.

10. ____ proposed double helix structure of DNA.

11. The membrane that encloses the cell is called_____.

12. The genetic material of a cell is _____.

13. DNA stands for _____.

14. _____ is a major source of chemical energy in the cell.

15. _____ are simple cells, generally found in plants and animals, and have true nuclei.

ANSWER CHOICES FOR QUESTIONS 1 THROUGH 15

The following choices are possible answers to some of the questions. If an answer is not found in the following choices, write your own answer.

a. Cytosine, Thymidine

b. Adenine, Guanine

c. ATGC

d. Hydrogen bond

e. Covalent bond

f. Oxygen bond

g. Two

h. Three

i. Four

j. Nucleotide

k. Polypeptide

l. ACGU

m. Nucleus

n. Cytoplasm

o. Nucleoli

p. Plasma membrane

q. Mitochondria

r. Endoplasmic reticulum

s. Gene

t. Watson and Crick

u. Watson, Crick, and Wilkins

v. Watson, Crick, Wilkins, and Franklin

    w. Eukaryotic

    x. Prokaryotic

16. How many genes are there in the human genome?

17. What do you mean by expression of genes?

18. What is the difference between the translation and the transcription process?

19. What is the difference between mRNA and tRNA?

Microarrays

Microarray technology allows measurement of the expression values of thousands of different DNA molecules at a given point in the life cycle of an organism or cell. The expression values of the DNA can be compared to learn about thousands of processes going on simultaneously in living organisms. This large-scale, high-throughput, interdisciplinary approach enabled by genomic technologies is rapidly becoming a driving force of biomedical research. Microarrays can be used to characterize specific conditions such as cancer (Alizadeh et al., 2000). Microarray technology is used to identify whether the expression profiles differ significantly between the various groups.

2.1 MICROARRAY TECHNOLOGY

A DNA array is an orderly arrangement of thousands of single strands of DNA of known sequence deposited on a nylon membrane or glass or plastic. The array of detection units (probes) can be uniquely synthesized on a rigid surface such as glass, or it can be presynthesized and attached to the array platform. The microarray platforms or fabrication can be classified in three main categories based on the way the arrays were produced and on the types of probes used. These three types of microarray platform are known as spotted cDNA arrays, spotted oligonucleotide arrays, and *in situ* oligonucleotide arrays.

In *in situ* probe synthesis, a set of oligonucleotide DNA probes is used to make a mask of photolithographic masks. The addresses on the glass surface are activated through the holes in the photolithographic mask and chemically coupled to bases' probes starting with ATGC. Then steps are repeated to add the second nucleotides to the glass surface. The process is repeated until a unique probe nucleotide of predefined length and a sequence of nucleotide is attached to each of the thousands of addresses on the glass plate. The number of probes per square centimeter on the glass surface varies from a few probes to hundreds of thousands. Since the presynthesized probes are manufactured commercially, the manufacturer controls the types of probes used in the glass slide (chip), and researchers cannot choose their own probes, hence limiting the use of the chip. But this limitation also makes the comparison between the results obtained by different labs easy, provided the chips are manufactured by the same manufacturer.

In spotted cDNA arrays, the expressed sequence tags (ESTs) or cDNA clones are robotically placed on a glass slide or spotted on a glass slide. The ESTs are single-pass, partial sequences of cDNA clones. The DNA cloning is obtained by selective amplification of a selected segment of the DNA. This method is used to create clones in a desired number for further analysis. Once a solution containing the ESTs is prepared, the robotic machine puts a small amount of the solution through the pins on the glass or other surface. In *in situ* fabrication, the length of the DNA sequence is limited, whereas in the spotted array technique, the length of the sequence can be of any size. Since the spotted cDNA can be from an organism about which the genome information may be limited or no information is available, the spotted cDNA can be very useful in studying genome-wide transcriptional profiling.

cDNA arrays have long sequences of cDNA and can use unknown sequences. The cDNA arrays are easier to analyze but create more variability in a data set. The oligonucleotide arrays use short sequences of known sequences but produce more reliable data. Analyzing the data obtained using oligonucleotide arrays is more difficult than analyzing the data obtained by using cDNA arrays.

In a basic two-color microarray experiment (Figure 2.1.1), two mRNA samples are obtained from samples under two different conditions. These conditions may be classified as healthy or disease conditions, before treatment or after treatment conditions, etc. The cDNA is prepared from these mRNA samples using the reverse transcription procedure. Then the two samples are labeled using two different fluorescent dyes. Generally, red dye is used for a disease sample, and green dye is used for a healthy sample. Two samples are hybridized together to the probe on the slides. These slides are placed under the scanner to excite corresponding fluorescent dye. The scanner runs two times, one time to excite red dyes and the other time to excite green dyes. The digital image is captured for measuring the intensity of each pixel. Theoretically, the intensity

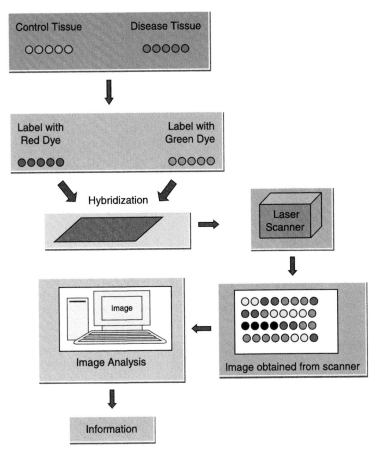

FIGURE 2.1.1

cDNA array processing. (See color insert).

of each spot is proportional to the amount of mRNA from the sample with the sequence matching the given spot. Two images obtained from two samples are overlapped, and artificial colors are provided by computer software to enhance the visual assessment. A gene expressed in the healthy sample will produce a green spot, while a gene expressed in the disease sample will produce a red spot, a gene expressed equally in two conditions will produce a yellow spot, and a gene not expressed at all under two conditions will produce a black spot (Figure 2.1.2, and Figure 2.1.3).

2.2 ISSUES IN MICROARRAY

Microarrays are relatively newer than some of the biotechnology available; therefore, they have some issues that need to be addressed.

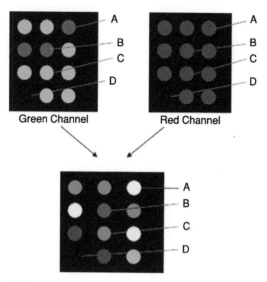

FIGURE 2.1.2
Microarray slides. (See color insert).

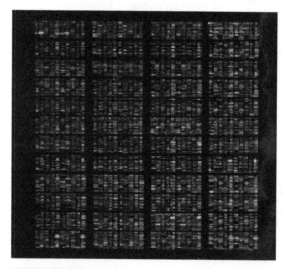

FIGURE 2.1.3
Scanned image obtained from affymetric machine. (See color insert).

Noise

One of the major difficulties in decoding gene expression data is related to the nature of the data. Even if an experiment is repeated twice with exactly the same material, machines, men, and conditions, it is possible that after scanning and image processing, many genes will have different quantitative expression values. Noise is introduced at every step of the microarray procedure from mRNA preparation, transcription, labeling, dyes, pin type, humidity, target volume, hybridization, background, inter-channel alignment, quantification, image analysis, etc. In general, the changes in the measured transcript values among different experiments are caused by both biological variations and experimental noise. To correctly interpret the gene expression microarray data, one needs to understand the sources of the experimental noise. There have been some advances in this area to reduce the noise in data sets. A detailed noise analysis for oligonucleotide-based microarray experiments involving reverse transcription, generation of labeled cDNA (target) through *in vitro* transcription, and hybridization of the target to the probe immobilized on the substrate was studied by Tu et al. (2002). Replication is also used to reduce noise in the data.

Normalization

The gene expression data is full of systematic differences across different data sets. To eliminate these systematic differences, a normalization technique is used. Once the data are normalized, it is possible to combine different experiment results. Many methods have been suggested to normalize the gene expression data (Park et al., 2003). Yang et al. (2002) described a number of normalization methods for dual-labeled microarrays, which included global normalization and locally weighted scatter plot smoothing. Kepler et al. (2002) extended the methods for global and intensity-dependent normalization methods. Wang et al. (2002) suggested an iterative normalization of cDNA microarray data for estimating normalized coefficients and identifying control genes. The main idea of normalization for dual-labeled arrays is to adjust for artificial differences in intensity of the two labels. Systematic variations cannot be controlled only by the normalization process, but the normalization process plays an important role in the primary microarray data analysis. Microarray data obtained by different normalization methods can significantly vary from each other and hence lead to different conclusions. The secondary stage of data analysis may also depend on the normalization procedure done initially.

Experimental Design

The experimental design allows a researcher to manipulate different input variables to see changes in the variables of interest. The design of the experiment depends on the purpose of the experiment. The purpose of a microarray experiment could be class prediction, class discovery, or class comparison (Simon and Dobbin, 2003). There are different types of experiments, such as *single-slide* cDNA microarray experiments, in which one compares transcript abundance in two samples, the red and green labeled samples hybridized to the same slide, and *multiple-slide* experiments comparing transcript abundance in two or more types of samples hybridized to different slides (Dudoit et al., 2002). In time-course experiments, the gene expression is observed as it changes over a period of time. There are several common designs available for microarray experiments, such as reference design, balanced-block design, loop design, and interwoven loop design. Each design has its own advantages and disadvantages. The most popular design is a reference design because of the ease with which it can be applied. More details about designs are provided in Chapter 10.

Sample Size and Number of Genes

In the classical testing of hypotheses, most of the available statistical tests can be applied for large samples. If the sample size is below five, most of the statistical tests will not produce true results. For example, t-tests work very well for a sample size of five or more. In the case of microarray experiments, the number of replications is generally very small due to the cost of conducting experiments. Microarrays produce thousands of variables from a small number of samples. It is common in microarray experiments to use three or fewer replicates (Breitling and Herzyk, 2005). If the sample size is small, most of the statistical methods cannot be applied directly without modification. Statistical approaches to microarray research are not yet as routine as they are in other fields of science. Statistical methods, suitable to microarrays, continue to be adapted and developed. The microarray data require new or modified techniques that can be applied in presence of large variables in small sample size (Nadon and Shoemaker, 2002). Due to immediate need for developing or modifying new methods to suit the microarray conditions, hundreds of methods have evolved in literature, which vary from simple fold-change approaches to testing for differential expression, to many more complex and computationally demanding techniques. It is seen that most of the methods can be regrouped and are in fact special cases of general approaches (Allison et al., 2006). Now efforts should be made to choose the statistical procedures that may suit the aim of the investigation and are efficient.

Other Issues

Quality of the array is another issue in microarrays. Some arrays that do not meet the standard quality may lead to the failure of a microarray experiment. The array quality assessment is addressed by several researchers (Carter et al., 2005; Imbeaud et al., 2005). The real expression level is the amount of the proteins produced by the gene and is not directly proportional to the amount of mRNA. Thus, expression measurements by microarray, which are based on the amount of the mRNA present, have been one of the big concerns. To overcome this problem, the microarray technique is modified to measure the amount of the proteins produced by the gene. Also, a new field is emerging called protein microarray or sometimes simply proteomics, which covers functional analysis of gene products, including large-scale identification or localization studies of proteins and interaction studies. More details are provided in Section 2.4.

2.3 MICROARRAY AND GENE EXPRESSION AND ITS USES

With DNA microarray technology, it is now possible to monitor expression levels of thousands of genes under a variety of conditions. There are two major application forms for the DNA microarray technology: (1) identification of sequence (gene/gene mutation); and (2) determination of expression level (abundance) of genes.

The DNA microchip is used to identify mutations in genes such as BRCA1 and BRCA2, which are responsible for cancer. To identify the mutation, one obtains a sample of DNA from the patient's blood as well as a control sample from a healthy person. The microarray chip is prepared in which the patient's DNA is labeled with the green dye and the control DNA is labeled with the red dye and allowed to hybridize. If the patient does not have a mutation for the gene, both the red and green samples will bind to the sequences on the chip. If the patient does possess a mutation, then the patient's DNA will not bind properly in the region where the mutation is located.

Microarrays are being applied increasingly in biological and medical research to address a wide range of problems, such as the classification of tumors or the gene expression response to different environmental conditions. An important and common question in microarray experiments is the identification of differentially expressed genes, i.e., genes whose expression levels are associated with response or covariate of interest. The covariates could be either polytomous (for example, treatment/control status, cell type, drug type) or continuous (for example, dose of a drug, time). The responses could be, for example, censored survival times or provide other clinical outcomes. The types of experiments include single-slide cDNA microarray experiments and multiple-slide experiments. In single-slide experiments, one compares transcript abundance (i.e., gene expression levels) in two mRNA samples, the red and green labeled mRNA samples hybridized to the same slide. In multiple-slide experiments comparing transcript abundance in two or more types of mRNA, samples are hybridized to different slides.

The most commonly used tools for identification of differentially expressed genes include qualitative observation (generally using the cluster analysis), heuristic rules, and model-based probabilistic analysis. The simplest heuristic approach is setting cutoffs for the gene expression changes over a background expression study. Iyer et al. (1999) sought genes whose expression changed by a factor of 2.20 or more in at least two of the experiments. DeRisi et al. (1997) looked for a two-fold induction of gene expression as compared to the baseline. Xiong et al. (2001) identified indicator genes based on classification errors by a feature wrapper (including linear discriminant analysis, logistic regression, and support vector machines). Although this approach is not based on specific data modeling assumptions, the results are affected by assumptions behind the specific classification methods used for scoring.

The probabilistic approaches applied to microarray analysis include a t-test based on a Bayesian estimate of variance among experiment replicates with a Gaussian model for expression measurements (Long et al., 2001) and hierarchical Bayesian modeling framework with Gaussian gene-independent models combined with a t-test (Baldi and Long, 2001). Newton et al. (2001) identified differentially expressed genes by posterior odds of change based on a hierarchical Gamma-Gamma-Bernoulli model for expression ratios. All these methods use either an arbitrarily selected cutoff or probabilistic inference based on a specific data model.

Most of the microarray data are not normally distributed, and often noise is present in the data (Hunter et al., 2001). It sometimes becomes difficult to have a test that can take care of noise as well as the rigid parametric condition of normality. We also want our tests to be applicable to all types of microarray data. In this context, we find nonparametric methods to be very useful. In earlier studies, the use of rank-transformed data in microarray analysis was advantageous (Raychaudhri et al., 2000; Tsodikov et al., 2002). A nonparametric version of a t-test was used by Dudoit et al. (2002). Tusher et al. (2001) used a significance analysis of microarray data, which used a test statistic similar to a t-test. A mixture model technique was used by Pan et al. (2001); it estimated the distribution of t-statistic-type scores using the normal mixture models. There are many other statistical procedures available to deal with the expression data.

2.4 PROTEOMICS

Protein microarrays are a new technology that has great potential and challenges (Hall et al., 2007). Proteomics is a complementary science to genomics that focuses on the gene products, whereas genomics focuses on genes. Protein microarrays are used for the identification and quantification of proteins, peptides, low molecular weight compounds, and oligosaccharides or DNA. These arrays are also used to study protein interactions. The idea and theoretical considerations of protein microarray were discussed by Ekins and Chu (1999). The protein microarray leads to quantification of all the proteins expressed in a cell and functional study of thousands of proteins at a given point of time. Protein-protein and protein-ligand (protein ligand is an atom, a molecule that can bind to a specific binding site on a protein) interactions are studied in the protein functional studies. Protein chips have a wider application, such as the identification of protein-protein interactions, protein-phospholipid interactions, small molecule targets, substrates of protein kinases, clinical diagnostics, and monitoring of disease states. The main aim of any biological study is to understand complex cellular systems and determine how the various components work together and are regulated. To achieve this aim, one needs to determine the biochemical activities of a protein, their relation to each other, the control proteins exercise themselves, and the effect of activities of proteins on each other. The protein microarray can be classified in three categories broadly based on their activities. These are known as analytical protein microarrays, functional protein microarrays, and reverse phase protein microarrays. The analytical protein microarrays are used to measure the level of proteins in a mixture, to measure the binding affinities in the protein mixture and specificities (Bertone and Snyder, 2005).

Analytical protein microarrays are very useful in studying the differential expression profiles and clinical diagnostics such as profiling responses to environmental stress and healthy versus disease tissues (Sreekumar et al., 2001).

Functional protein microarrays contain arrays of full-length functional proteins or protein domains. With these protein chips, biochemical activities of an entire proteome can be studied in a single experiment and can be used to study many protein interactions, such as protein DNA, protein-protein, protein-RNA, protein-cDNA, etc. (Hall et al., 2004).

In the reverse phase protein microarray (RPA), sample lysate (contents released from a lysed cell) are prepared by isolating the cells from the tissues under consideration, and cells are isolated from various tissues of interest and are lysed (*lysis* refers to the death of a cell by breaking of the cellular membrane). The lysate is arrayed on a glass slide or other type of slide. These slides are then used for further experimentation. RPAs can be used to detect the post-transcriptional changes in the proteins that may have occurred due to a disease. This helps in treating the dysfunctional protein pathways in the cell and hence can treat the disease successfully.

Proteome chip technology is used to probe entire sets of proteins for a specific cell or its function. This extremely valuable technology has vast applications and may be useful for discovering different interactions between proteins, their functions, and the effect of their biochemical activities.

EXERCISE 2

A patient with cancer, A, has the following gene expression pattern, as shown in Table 2.4.1. The expressed gene is shown by a black circle, whereas the unexpressed gene is shown by a white circle.

Table 2.4.1 Expression Pattern of a Patient with Cancer, A

Gene #	1	2	3	4	5	6	7	8	9	10
Cancer A	●	○	●	○	●	○	●	○	●	●

This patient responds very well to a newly developed drug, A. Several other patients had different types of cancers with gene expression given in Table 2.4.2. Find out which type(s) of cancer will respond very well to drug A.

Table 2.4.2 Expression Pattern of Patients having Different Types of Cancers

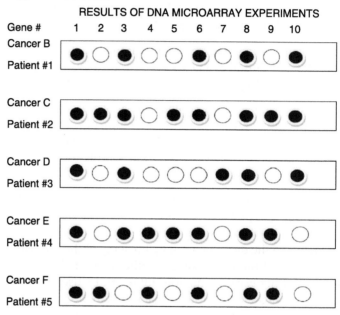

Probability and Statistical Theory

CONTENTS

Mathematical theory of probability was developed by two famous French mathematicians, Blaise Pascal and Pierre de Fermat, in the 16th century and later developed by Jakob Bernoulli (1654–1705), Abraham de Moivre (1667–1754), and Pierre de Laplace (1749–1827). A gambler asked Pascal whether it would be profitable to bet on getting at least one pair of sixes in a throw of two dice 24 times. This problem led to an exchange of letters between Pascal and Fermat in which fundamental principles of probability theory were formulated. Since then, the study of probability has become very popular, and its applications have grown rapidly. In bioinformatics, probability is applied extensively. Most of the mathematical models are probabilistic in nature.

3.1 THEORY OF PROBABILITY

If an experiment is repeated under essentially homogeneous and similar conditions, it will lead to two types of outcomes:

(i) The outcome is unique or deterministic in nature. For example, when two molecules of hydrogen are mixed with a molecule of oxygen, water is obtained. Similarly, the velocity (v) of an object can be found using the equation, $v = u + at$ where t is the time, a is the acceleration, and u is the initial velocity. When this type of experiment is repeated under homogeneous conditions, then the outcome will always be the same.

(ii) The outcome is not the same. In such cases, if an experiment is repeated under homogeneous conditions, it will lead to one of several outcomes; hence, the outcome cannot be predicted accurately. For example, when an unbiased coin is tossed, the outcome can be either heads or tails. Thus, the outcome cannot be predicted accurately. Similarly, if an unbiased die is tossed, the outcome can be either 1, 2, 3, 4, 5, or 6, and it is not possible to predict the outcome with certainty. Similarly, in life-testing experiments, it is not possible to predict how long equipment will last. Also, in the case of gene expressions, it cannot be predicted with certainty that each time the same number of genes will be expressed if the experiment is repeated under similar conditions.

In situation (ii), where the outcome is uncertain, the chance or probability measures quantitative certainty.

A statistical experiment, which gives one of the several outcomes when repeated essentially under similar conditions, is known as a *trial,* and outcomes of the trial are known as *events*. The set of all possible outcomes of a statistical experiment is called the *sample space* and is represented by the symbol S.

■ Example 1

In the case of tossing a coin three times, the *trial* is tossing the coin, which is repeated three times, and *events* are heads or tails obtained in three repetitions. ■

■ Example 2

In microarray experiments, gene prediction is highly important.

It is known that it is not possible to predict a gene expression with certainty. In such situations, the microarray experiment becomes a *trial,* and the gene expression value becomes an *event*.

The total number of possible outcomes in any trial is called an *exhaustive event*. ∎

■ Example 3

In the case of tossing a coin three times, each trial results in two outcomes—namely, heads or tails. Since a coin is tossed three times, the total number of possible outcomes is $2^3 = 8$.

Favorable events are those events in a trial that favor the happening of the events. ∎

■ Example 4

If we throw a die and try to get an even number, then the events that are in favor of getting an even number are $\{2, 4, 6\}$. ∎

■ Example 5

A pair of alleles governs the coat color in a guinea pig. A dominant allele B leads to black color, and its recessive allele b leads to a white coat color. The F_1 cross produces the exhaustive number of cases $\{BB, Bb, bB, bb\}$. The favorable event for getting a white guinea pig is $\{bb\}$.

Events are said to be *mutually exclusive* if the happening of any one of them precludes the happening of all others. Thus, only one event can occur at a time. ∎

■ Example 6

Consider example 5 again. If event A is getting a black guinea pig in F_1 cross and event B is getting a white guinea pig in the same F_1 cross, then events A and B are mutually exclusive because only one of the two events can occur at a time (in F_1 cross, only a black guinea pig or white guinea pig is possible). ∎

■ Example 7

A pair of alleles governs flower color in flowering plants such as snapdragons. A first cross between pure-breeding red-flowered plants (R) and pure-breeding white-flowered plants (W) results in pink plants (RW). The exhaustive number of events is $\{RR, RW, WR, WW\}$. If event A is getting a pink flower, event B is getting a white flower, and event C is getting a red flower in the first cross, then events A, B, and C are mutually exclusive events since only one of the possible colors can occur in the F_1 cross.

Several events are said to be *independent* if the happening (or nonhappening) of an event is not affected by the happening of any event or other events of a trial.   ∎

∎ Example 8

Consider the following polypeptide chain:

$$\dots 5'\text{AUG GAA GCA UCA CCC UAG } 3' \dots$$

If A is an event getting a codon starting with U, and B is an event of getting a codon starting with C in the polypeptide chain, then the two events A and B are independent because getting a codon starting with U is not going to affect the occurrence of a codon starting with C or vice versa.   ∎

3.2 MATHEMATICAL OR CLASSICAL PROBABILITY

Definition *Let E be an event of interest in trial T, which results in n exhaustive, mutually exclusive, and equally likely cases. Let m events out of n events be favorable to the happening of the event E.*

The probability p of E happening is given by

$$p = \frac{\text{Favorable number of cases}}{\text{Exhaustive number of cases}} = \frac{m}{n}.$$

If the number of cases favorable to the "nonhappening" of event E are $(n - m)$*, then the probability that the event E will not happen is*

$$q = \frac{n - m}{n} = 1 - \frac{m}{n} = 1 - p \Rightarrow p + q = 1.$$

The probability is always non-negative and $0 < p < 1$*, and* $0 < q < 1$*.*

∎ Example 1

In example 7, the first cross between pure-breeding red-flowered plants (R) and pure-breeding white-flowered plants (W) results in pink plants (RW). The exhaustive number of events is {RR, RW, WR, WW}. If event A is getting a pink flower, event B is getting a white flower, and event C is getting a red flower in the first cross, then find the probability of getting

    a. a pink flower.

    b. a white flower.

    c. a red flower.

Solution

Exhaustive number of cases {RR, RW, WR, WW} = 4

    a. Number of favorable cases for getting a pink flower {RW, WR} = 2. Then the probability of getting a pink flower is

$$p = \frac{Favorable\ number\ of\ cases}{Exhaustive\ number\ of\ cases} = \frac{2}{4}.$$

    b. Number of favorable cases for getting a white flower {WW} = 1. Then the probability of getting a white flower is

$$p = \frac{Favorable\ number\ of\ cases}{Exhaustive\ number\ of\ cases} = \frac{1}{4}.$$

    c. Number of favorable cases for getting a red flower {RR} = 1. Then the probability of getting a red flower is

$$p = \frac{Favorable\ number\ of\ cases}{Exhaustive\ number\ of\ cases} = \frac{1}{4}.$$

■

■ Example 2

A bag contains four red, three white, and eight black balls. What is the probability that two balls drawn are red and black?

Solution

Total number of balls = 4 + 3 + 8 = 15.

The exhaustive number of cases = out of a total of 15 balls, two balls (red and black) can be drawn in $^{15}C_2$ ways = $^{15}C_2$ =

$$\frac{15!}{2!13!} = \frac{15 \times 14}{2} = 105.$$

Number of favorable cases = out of two red balls, one ball can be drawn in $^4C_1$, and out of eight black balls, one ball can be drawn in $^8C_1$ ways = $^4C_1 \times {}^8C_1 = 4 \times 8 = 32$.

Then the probability of drawing a red and black ball is

$$p = \frac{Favorable\ number\ of\ cases}{Exhaustive\ number\ of\ cases} = \frac{32}{105}.$$

Following is the R-code for example 2:

```
#Example 2
> num<-choose(4,1)*choose(8,1)
```

```
> deno<-choose(15,2)
> #p = Probability of drawing a red and a black ball
> p = num/deno
> cat("P(Drawing a red and a black ball) = ",p,fill = T)
P(Drawing a red and a black ball) = 0.3047619
```
■

■ Example 3

In a microarray experiment, out of 10,000 spots, 8,900 spots are circular, and 1100 spots are uniform. What is the probability of getting a uniform spot when a spot is chosen randomly?

Solution

Favorable number of cases $= 1100$.

Total number of cases $= 10,000$.

Then the probability of getting a uniform spot is

$$p = \frac{Favorable\ number\ of\ cases}{Exhaustive\ number\ of\ cases} = \frac{1100}{10,000} = 0.11.$$

Following is the R-code for example 3:

```
#Example 3
> # Num = Favorable number of cases
> # Deno = Exhaustive number of cases
> # p = probability of getting a uniform spot
> Num<-1100
> Deno<-10000
> p = Num/Deno
> cat("P(Getting a uniform spot) = ",p,fill = T)
P(Getting a uniform spot) = 0.11
```
■

3.3 SETS

A set is a well-defined collection *or* aggregate of all possible objects having given properties and specified according to a well-defined rule. The objects in the set are known as elements of the set. Generally a set is denoted by capital letters. If any object z is an element of the set A, then the relationship between z and A is denoted by $z \in A$. If z is not a member of A, then the relationship is denoted by $z \notin A$. If each and every element of the set A belongs to the set B, that is, if $z \in A => z \in B$, then A is said to be a subset of B (or B is a superset of A). It is therefore denoted by either $A \subseteq B$ (A is contained in B) or $B \supseteq A$ (B contains A). Two sets A and B are said to be *equal* or *identical* if $A \subseteq B$ and $B \subseteq A$ and *write* $A = B$ or $B = A$. A *null* or an *empty* set is one that does not contain any element at all, and it is denoted by Φ.

3.3.1 Operations on Sets

The *union* of two given sets A and B, denoted by $A \cup B$, is defined as a set consisting of all these points, which belongs to either A or B or both. It is denoted as

$$A \cup B = \{z : z \in A \text{ or } z \in B\}.$$

For sets of n elements, the union is denoted by

$$\bigcap_{i=1}^{n} A_i = \{z : z \in A_i \text{ for at least one } i = 1, 2, \ldots, n\}.$$

Similarly, the *intersection* of two sets A and B, represented by $A \cap B$, is defined as a set consisting of those elements that belong to both A and B.

For sets A and B, the intersection is denoted by

$$A \cap B = \{z : z \in A \text{ and } z \in B\}.$$

Similarly, for n sets A_i, $i = 1, 2, \ldots, n$,

$$\bigcap_{i=1}^{n} A_i = \{z : z \in A_i \text{ for all } i = 1, 2, \ldots, n\}.$$

On the other hand, if A and B have no common point, i.e., $A \cap B = \Phi$, then the sets A and B are said to be *disjoint*, or *mutually exclusive*.

The *complement* or *negative* of any set A, denoted by $\overline{A}$ or A^c or A', is a set containing all elements of the universal set S that are not elements of A, i.e., $\overline{A} = S - A$.

■ Example 1

If $S = \{1, 2, 3, 4, 5, 8, 10, 12\}$, $A = \{1, 2, 3, 8, 10\}$, and $B = \{2, 4, 8, 12\}$, then find

    a. $A \cup B$.

    b. $A \cap B$.

    c. $\overline{A}$.

Solution

    a. $A \cup B = \{1, 2, 3, 4, 8, 10, 12\}$.

    b. $A \cap B = \{2, 8\}$.

    c. $\overline{A} = \{4, 5, 12\}$.

The following R-code performs the operations needed in example 1:

```
#Example 1
> setS<-c(1,2,3,4,5,8,10,12)
> SetA<-c(1,2,3,8,10)
SetB<-c(2,4,8,12)
> union(SetA,SetB)
[1] 1 2 3 8 10 4 12
> intersect(SetA,SetB)
[1] 2 8
```

■

3.3.2 Properties of Sets

If A, B, and C are the subsets of a universal set S, then following are some of the useful properties.

Commutative Law:

$$A \cup B = B \cup A, A \cap B = B \cap A.$$

Associative Law:

$$(A \cup B) \cup C = A \cup (B \cup C);$$
$$(A \cap B) \cap C = A \cap (B \cap C).$$

Distributive Law:

$$A \cap (B \cup C) = (A \cap B) \cup (A \cap C);$$
$$A \cup (B \cap C) = (A \cup B) \cap (A \cup C).$$

Complementary Law:

$$A \cup \overline{A} = S; A \cap \overline{A} = \Phi;$$
$$A \cup S = S; A \cap S = A;$$
$$A \cup \Phi = A; A \cap \Phi = \Phi.$$

Difference Law:

$$A - B = A \cap B;$$
$$A - B = A - (A \cap B) = (A \cup B) - B;$$
$$A - (B - C) = (A - B) \cup (A - C);$$
$$(A \cup B) - C = (A - C) \cup (B - C);$$
$$A - (B \cup C) = (A - B) \cap (A - C);$$
$$(A \cap B) \cup (A - B) = A, (A \cap B) \cap (A - B) = \Phi.$$

De-Morgan's Law—Dualization Law:

$$(\overline{A \cup B}) = \overline{A} \cap \overline{B}, \quad (\overline{A \cap B}) = \overline{A} \cup \overline{B}.$$

More generally,

$$\left(\overline{\bigcup_{i=1}^{n} A_i} \right) = \bigcap_{i=1}^{n} \overline{A_i} \quad \text{and} \quad \overline{\bigcap_{i=1}^{n} A_i} = \bigcup_{i=1}^{n} \overline{A_i}.$$

Involution Law:

$$\overline{\overline{A}} = A.$$

Idempotency Law:

$$A \cup A = A.$$

In mathematical terms, probability can be defined as a function that assigns a non-negative real number to every event A, represented by $P(A)$, on a given sample space S.

The following properties *or* axioms hold for $P(A)$.

1. For each $A \in S$, $P(A)$ is defined, is real, and $P(A) \geq 0$.

2. $P(S) = 1$.

3. If $\{A_n\}$ is any finite or infinite sequence of disjoint events in S, then

$$P \left(\bigcup_{i=1}^{n} A_i \right) = \sum_{i=1}^{n} P(A_i).$$

The preceding three axioms are termed as the axiom of positiveness, certainty, and union, respectively.

3.4 COMBINATORICS

Combinatorics is a branch of mathematics concerned with counting, arrangements, and ordering. The following sections describe four major theorems and rules that have wide applications in solving problems.

Multiplication Rule

If experiment A can be performed in m different ways and experiment B can be performed in n different ways, the sequence of experiments A and B can be performed in mn different ways.

■ Example 1

A biologist would like to see the effect of temperature, dye, and glass slide on microarray images. If the biologist uses two different temperatures, two dyes, and three glass slides, how many different experiments does the biologist need to perform to observe each temperature-dye-glass slide exactly twice?

Solution

Number of different experiments

$$= (\text{two different temperatures} \cdot \text{two dyes} \cdot \text{three glass slides}) \cdot 2$$

$$= (2 \cdot 2 \cdot 3) \cdot 2$$

$$= 24.$$

■

Counting Permutation (When the Objects Are All Distinct and Order Matters)

The number of permutations of selecting k objects from a set of n distinct elements, denoted by $_nP_k$, is given by

$$_nP_k = n(n-1)(n-2)\ldots(n-(k-1)) = \frac{n!}{(n-k)!}.$$

■ Example 2

A microbiologist decides to perform an operation on three bacteria, five viruses, and three amphibians. The microbiologist will perform the operations on three subjects in order. In how many ways can the microbiologist choose three subjects?

Solution

Here $n = 11$ and $k = 3$.

$$\text{Number of permutations} = \frac{11!}{(11-3)!} = 990.$$

Following is the R-code for example 2.

```
#Example 2
> permutations(n = 11,r = 3)
#Note: it will generate list of permuted values also.
.....
[990,]   11   10   9
```

■

Counting Permutations (When the Objects Are Not All Distinct and Order Matters)

The number of ways to arrange n objects, out of which n_1 objects are of the first kind, n_2 objects are of the second kind, ..., and n_k objects are of kth kind, is given by

$$\frac{n!}{n_1! n_2! \cdots n_k!}, \quad \text{where } \sum_{i=1}^{k} n_i = n.$$

■ Example 3

In how many ways can a biologist arrange three flies, two frogs, and five mice on an operation table when order in which subjects are arranged on the operation table matters?

Solution

Here number of flies $= n_1 = 3$,

$$\text{Number of frogs} = n_2 = 2,$$
$$\text{Number of mice} = n_3 = 5.$$

Then the number of ways in which three kinds of objects can be arranged

$$= \frac{10!}{3! 2! 5!} = 2520.$$

Following is the R-code for example 3.

```
> #n1 = number of flies, n2 = number of frogs, n3 = number of mice
> #n = n1 + n2 + n3
> n1<-3
> n2<-2
> n3<-5
> n = n1 + n2 + n3
> p = factorial(n)/(factorial(n1)*factorial(n2)*factorial(n3))
> cat("number of ways in which three kinds of objects can be arranged = ", p,fill = T)
number of ways in which three kinds of objects can be arranged = 2520   ■
```

Counting Combinations (When the Objects Are All Distinct and Order Does Not Matter)

The number of ways to select k objects from n distinct objects is given by

$$\binom{n}{k} = \frac{n!}{k!(n-k)!}.$$

■ Example 4

There are 20 mice in a cage. An experimenter would like to take out five mice randomly from this cage. In how many ways can the experimenter draw five mice from the cage?

Solution

Here $n = 20$ and $k = 5$.

Number of ways in which the experimenter can draw five mice

$$= \binom{20}{5} = \frac{20!}{5!15!} = 15504.$$

Following is the R-code for example 4.

```
> #n = number of mice in a cage, k = number of mice drawn from the cage
> #c = Number of ways in which k mice can be drawn
> n<-20
> k<-5
> c = choose(n,k)
> cat("Number of ways in which 5 mice can be drawn from the cage = ",c,fill = T)
Number of ways in which 5 mice can be drawn from the cage = 15504       ■
```

3.5 LAWS OF PROBABILITY

The following *laws of probability* hold good:

1. Probability of the impossible event is zero, i.e., $P(\Phi) = 0$.

2. Probability of the complementary event is given by
$$P(\overline{A}) = 1 - P(A).$$

3. For any two events A and B,
$$P(\overline{A} \cup B) = P(B) - P(A \cap B).$$

4. Probability of the union of any two events A and B is given by
$$P(A \cap \overline{B}) = P(A) - P(A \cap B).$$

5. If $B \subset A$, then
   i. $P(A \cap \overline{B}) = P(A) - P(B)$,
   ii. $P(B) \leq P(A)$.

6. Law of Addition of Probabilities
   If A and B are any two events (subsets of sample space S) and are not disjoint, then
$$P(A \cup B) = P(A) + P(B) - P(A \cap B).$$

7. Extension of General Law of Addition of Probabilities
For events $A_1, A_2, \ldots, A_n$, we have

$$P\left(\bigcup_{i=1}^{n} A_i\right) = \sum_{i=1}^{n} P(A_i) - \sum_{1 \leq i \leq j \leq n} \sum P(A_i \cap A_j)$$

$$+ \sum_{1 \leq i \leq j \leq k \leq n} \sum\sum P(A_i \cap A_j \cap A_k)$$

$$+ \cdots + (-1)^{n-1} P(A_1 \cap A_2 \cap \ldots \cap A_n).$$

8. Boole's Inequality
For n events, $A_1, A_2, \ldots, A_n$, we have

a. $P\left(\bigcap_{i=1}^{n} A_i\right) \geq \sum_{i=1}^{n} P(A_i).$

b. $P\left(\bigcup_{i=1}^{n} A_i\right) \leq \sum_{i=1}^{n} P(A_i).$

9. Multiplication Law of Probability and Conditional Probability
For n events A and B, we have

$$P(A \cap B) = P(A)P(B|A), P(A) > 0$$
$$= P(B)P(A|B), P(B) > 0.$$

where $P(B|A)$ is the conditional probability occurrence of B when the event A has already occurred. $P(A|B)$ is the conditional probability of an event A given that B has already occurred.

10. Given n independent events $A_i (i = 1, 2, 3, \ldots, n)$ with respective probabilities of occurrence p_i, the probability of occurrence of at least one of them is given by

$$P(A_1 \cup A_2 \cup \ldots \cup A_n)$$
$$= 1 - P(\bar{A}_1 \cap \bar{A}_2 \cap \bar{A}_3 \ldots \cap \bar{A}_n)$$
$$= 1 - P(\bar{A}_1)P(\bar{A}_2)P(\bar{A}_3) \ldots P(\bar{A}_n)$$
$$= 1 - [(1 - p_1)(1 - p_2) \ldots (1 - p_n)]$$
$$= \left[\sum_{i=1}^{n} p_i - \sum_{\substack{i,j=1 \\ i<j}}^{n} \sum (p_i p_j) + \sum_{\substack{i,j,k=1 \\ i<j<k}}^{n} \sum\sum p_i p_j p_k - \cdots \right.$$

$$\left. + (-1)^{n-1}(p_1 p_2 p_3) \right].$$

11. For any three events A, B, and C,

$$P(A \cup B|C) = P(A|C) + P(B|C) - P(A \cap B|C).$$

12. For any two events A and B defined on a sample space S with $P(B) > 0$
  i. $P(A|B) = \frac{P(A \cap B)}{P(B)} \geq 0$
  ii. $P(S|B) = \frac{P(S \cap B)}{P(B)} = \frac{P(B)}{P(B)} = 1$
  iii. If $\{A\}$ is any finite or infinite sequence of disjoint events, then

$$P\left(\bigcup_n A_n|B\right) = \frac{P\left(\left(\bigcup_n A_n\right) \cap B\right)}{P(B)} = \frac{P\left(\bigcup_n A_n B\right)}{P(B)}$$

$$= \frac{\sum_n P(A_n B)}{P(B)} = \sum_n \left(\frac{P(A_n B)}{P(B)}\right) = \sum_n P(A_n|B).$$

13. Let events A, B, and C be defined on the sample space S such that $B \subset C$ and $P(A) > 0$; then

$$P(B/A) \leq P(C/A).$$

14. Independent Events

Two events A and B are said to be independent if the conditional probability of B given A, i.e., $P(B|A)$, is equal to the unconditional probability of B, i.e.,

$$P(B|A) = P(B).$$

Also $P(B|A) = P(B)$ and $P(A|B) = P(A)$ when A and B are independent. Thus, the events A and B are independent if

$$P(A \cap B) = P(A)P(B).$$

15. Pairwise Independent Events

A set of events $A_1, A_2, \ldots, A_n$ is said to be pairwise independent if

$$P(A_i \cap A_j) = P(A_i)P(A_j) \ \forall \ i \neq j.$$

16. Mutual Independence of n Events

The events in sample space S are said to be mutually independent if the probability of the simultaneous occurrence of any finite number of them is equal to the product of their separate probabilities.

If $A_1, A_2, \ldots, A_n$ are n events, then for their mutual independence, we should have
  i. $P(A_i \cap A_j) = P(A_i)P(A_j)$, $(i \neq j, i, j = 1, 2, \ldots, n)$,
  ii. $P(A_i \cap A_j \cap A_k) = P(A_i)P(A_j)P(A_k)$, $(i \neq j \neq k, i, j, k = 1, 2, \ldots, n)$.

The probability of simultaneous occurrence of other events can be generalized similarly. For example, for three events A_1, A_2, A_3, the conditions for mutual independence are given by

$$P(A_1 \cap A_2) = P(A_1)P(A_2),$$
$$P(A_2 \cap A_3) = P(A_2)P(A_3),$$
$$P(A_1 \cap A_3) = P(A_1)P(A_3),$$

and

$$P(A_1 \cap A_2 \cap A_3) = P(A_1)P(A_2)P(A_3).$$

It is of interest to note that the pairwise independence does not imply mutual independence.

17. If A and $\overline{B}$ are independent events, then A and B are also independent events.

18. If A and B are independent events, then $\overline{A}$ and $\overline{B}$ are also independent events.

19. For any two events A and B,

$$P(A \cap B) \leq P(A) \leq P(A \cup B) \leq P(A) + P(B).$$

20. Bayes' Theorem
If $A_1, A_2, \ldots, A_n$ are mutually independent events with $P(A_i) \neq 0, i = 1, 2, \ldots, n$, then for any event $C, C \subset \overset{n}{\underset{i=1}{\cup}} A_i$, such that $P(C) > 0$,

$$P(A_i|C) = \frac{P(A_i)P(C|A_i)}{\sum\limits_{i=1}^{n} P(A_i)P(C|A_i)}.$$

Bayes' theorem can also be expressed as follows:
If A and B are two events, then

$$P(A|B) = \frac{P(A)P(B|A)}{P(A)P(B|A) + P(A')P(B|A')}.$$

■ Example 1

In a microarray experiment, a spot is acceptable if it is either perfectly circular or perfectly uniform or both. In an experiment, a researcher found that there are a total of 20,000 spots. Out of these 20,000 spots, 15,500 are circular, 9,000 are uniform, and 5,000 spots are both circular and uniform. If a spot is chosen randomly, find the probability that the selected spot will be acceptable.

Solution

P(circular spot) $= 15500/20000 = 0.775$.

P(uniform spot) $= 9000/20000 = 0.45$.

P(circular and uniform spot) $= 5000/20000 = 0.25$.

P(a randomly chosen spot is acceptable)

$\quad = P$(circular) $+ P$(uniform spot) $- P$(circular and uniform)

(because of law 5)

$\quad = 0.775 + 0.45 - 0.25 = 0.975$.

Following is the R-code for example 1:

```
> #n = Total number of spots
> ># uc = Number of uniform and circular spots
> # c = Number of circular spots
> # u = Number of uniform spots
> n<-20000
> c<-15500
> u<-9000
> uc<-5000
> p = c/n + u/n-uc/n
> cat("Probability that a randomly chosen spot is acceptable = ", p, fill = T)
Probability that a randomly chosen spot is acceptable = 0.975
```

∎

■ Example 2

In a microarray experiment, a dye R will be labeled into target cDNA with a probability of 0.95. The probability that a signal will be read from a spot (labeled and hybridized) is 0.85. What is the probability that a labeled target cDNA hybridizes on a spot?

Solution

Let A be the event that the target cDNA will be labeled with the dye R. Let B be the event that the target cDNA will hybridize with the labeled dye R. If it is given that $P(A) = 0.95$, and $P(A \cap B) = 0.85$, then the probability that the labeled target cDNA hybridizes on a spot is given by

$$P(B|A) = \frac{P(A \cap B)}{P(A)} = \frac{0.85}{0.95} = 0.89 \quad \text{(because of law 8)}.$$

Following is the R-code for example 2:

```
#Example 2
> # A is the event that the target cDNA will be labeled with the dye R
> #B is the event that the target cDNA will hybridize with labeled dye R
> PA<-0.95
> PAintersectionB<-0.85
> PBgivenA<-PAintersectionB/PA
> cat("Probability that the labeled target cDNA hybridizes on a spot = ", PBgivenA, fill = T)
Probability that the labeled target cDNA hybridizes on a spot = 0.8947368      ■
```

■ Example 3

In a microarray experiment, the probability that a target cDNA is dyed with dye G is 0.8. The probability that the cDNA will hybridize on a given spot is 0.85. The probability of getting a signal from a spot is 0.9. Can hybridization and labeling of the target be considered independent?

Solution

It is given that $P(\text{target is dyed with dye } G) = 0.8$.

$P(\text{cDNA hybridizes on a given spot}) = 0.85$.

$P(\text{signal is obtained from a spot}) = 0.9$.

If labeling of the target and hybridization are independent, then we must have

$P(\text{signal is obtained from a spot})$

$= P(\text{target is labeled with the dye } G) \cdot P(\text{target is hybridized})$
$$\text{(because of law 13).}$$

Substituting the probabilities, the equation becomes

$0.9 \neq 0.8 \cdot 0.85$.

Thus, hybridization and labeling of the target are not independent.

Following is the R-code for example 3:

```
> # PA is the probability that target is dyed with dye G
> # PB is the probability that cDNA hybridizes on a given spot
> # PS is the probability that signal is obtained from a spot
> PA<-0.8
> PB<-0.85
```

```
> PS<-0.9
> PS1<-PA*PB
> if (PS! = PS1) print("Hybridization and labeling of the target are NOT independent")
else print ("Hybridization and labeling of the target are independent")
[1] "Hybridization and labeling of the target are NOT independent"
```
■

■ Example 4

Consider two genes 3030 and 1039, which are working independently of each other. The probability that gene 3030 will express is 0.55, and the probability that gene 1039 will express is 0.85 in a given condition. What is the probability that both genes 3030 and 1039 will express at the same time in a given condition?

Solution

$P(\text{gene 3030 expresses}) = 0.55$.

$P(\text{gene 1039 expresses}) = 0.85$.

Since both genes 3030 and 1039 are independent of each other, therefore,

$P(\text{both genes 3030 and 1039 express at the same time})$

$\quad = P(\text{gene 3030 expresses}) \cdot P(\text{gene 1039 expresses})$

(because of law 13)

$\quad = 0.55 \cdot 0.85 = 0.4675$.

The following R-code performs the probability operation needed in example 4:

```
># PA is the probability that gene 3030 expresses
> # PB is the probability that gene 1039 expresses
> # PS is the probability that both genes 3030 and 1039 express at the same time
> PA<-0.55
> PB<-0.85
> PS<-PA*PB
> cat("Probability that both genes 3030 and 1039 express at the same time
+ = ", PS, fill = T)
Probability that both genes 3030 and 1039 express at the same time
= 0.4675
```
■

■ Example 5

Three microarray experiments $E_1, E_2,$ and E_3 were conducted. After hybridization, an image of the array with hybridized fluorescent dyes (red and green) is acquired in each experiment. The DNA microchip is divided into several grids, and each grid is of one square microcentimeter. In experiment E_1, it was observed that on a randomly selected grid there were two red spots, seven green spots, and five black spots. In experiment E_2, there were three red spots, seven green spots, and three black spots in a randomly selected grid, while in experiment E_3, it was observed that there were four red spots, eight green spots, and three black spot on a randomly selected grid. A researcher chose a grid randomly, and two spots were chosen randomly from the selected grid. The two chosen spots happened to be red and black. What is the probability that these two chosen spots came from

a. experiment E_3?

b. experiment E_1?

Solution

Since all three grids have equal probability of selection,

$$P(E_1) = P(E_2) = P(E_3) = \frac{1}{3}.$$

Let A be the event that two spots chosen are red and black. Then

$$P(A|E_1) = \frac{^2C_1 \times {}^5C_1}{^{14}C_2} = \frac{10}{91} = 0.109.$$

$$P(A|E_2) = \frac{^3C_1 \times {}^3C_1}{^{13}C_2} = \frac{9}{78} = 0.115.$$

$$P(A|E_3) = \frac{^4C_1 \times {}^3C_1}{^{15}C_2} = \frac{12}{105} = 0.114.$$

a. Using Bayes' theorem, we get

$$P(E_3|A) = \frac{P(E_3)P(A|E_3)}{\sum\limits_{i=1}^{3} P(E_i)P(A|E_i)}$$

$$= \frac{\frac{1}{3} \times 0.114}{\frac{1}{3} \times 0.109 + \frac{1}{3} \times 0.115 + \frac{1}{3} \times 0.114} = \frac{0.038}{0.112} = 0.339.$$

b. Using Bayes' theorem, we find

$$P(E_1|A) = \frac{P(E_1)P(A|E_1)}{\sum\limits_{i=1}^{3} P(E_i)P(A|E_i)}$$

$$= \frac{\frac{1}{3} \times 0.109}{\frac{1}{3} \times 0.109 + \frac{1}{3} \times 0.115 + \frac{1}{3} \times 0.114} = \frac{0.0363}{0.112} = 0.324.$$

The following R-code is developed for doing computations for example 5:

```
> #Example 5
>
> #Three microarray experiments: E1, E2, and E3
> # Probability of choosing E1, E2, E3 are PE1, PE2, PE3 respectively
> # Red Spot = R, Green Spot = G, Black Spot = B
> #A is the event that two spots chosen are red and black
>
> PE1<-1/3
> PE2<-1/3
> PE3<-1/3
> # In E1: two red spots, seven green spots and five black spots, E1N: Total number of
Spots
>
> E1R<-2
> E1G<-7
> E1B<-5
> E1N<-E1R + E1G + E1B
> E1N
[1] 14
>
> # In E2: three red spots, nine green spots and three black spots, E2N: Total number of
Spots
>
> E2R<-3
> E2G<-7
> E2B<-3
> E2N<-E2R + E2G + E2B
> E2N
[1] 13
>
> # In E3: four red spots, eight green spots, and three black spots, E3N: Total number of
Spots
> E3R<-4
> E3G<-8
> E3B<-3
> E3N<-E3R + E3G + E3B
> E3N
[1] 15
>
> # PAgivenE1 = P(A|E1), PAgivenE2 = P(A|E2, PAgivenE3 = P(A|E3)
>
> PAgivenE1<-(choose(E1R,I)*choose(E1B,1))/choose(E1N,2)
>
```

```
> cat("P(A|E1) = ", PAgivenE1, fill = T)
P(A|E1) = 0.1098901
>
> PAgivenE2<-(choose(E2R,1)*choose(E2B,1))/choose(E2N,2)
>
> cat("P(A|E2) = ", PAgivenE2, fill = T)
P(A|E2) = 0.1153846
>
> PAgivenE3<-(choose(E3R,1)*choose(E3B,1))/choose(E3N,2)
>
> cat("P(A|E3) = ", PAgivenE3, fill = T)
P(A|E3) = 0.1142857
>
> # Deno1 = Sum(P(E_i)*P(A|E_i)), i = 1,2,3
>
> Deno1<-(PE1*PAgivenE1 + PE2*PAgivenE2 + PE3*PAgivenE3)
>
> cat("Sum(P(E_i)*P(A|E_i)), i = 1,2,3
+ = ", Deno1, fill = T)
Sum (P(E_i)*P(A|E_i)), i = 1,2,3
= 0.1131868
>
> #PE1givenA = P(E1|A), PE2givenA = P(E2|A), PE3givenA = P(E3|A)
>
> PE1givenA<-PE1*PAgivenE1/Deno1
>
> cat("P(E1|A) = ", PE1givenA, fill = T)
P(E1|A) = 0.3236246
>
> PE2givenA<-PE2*PAgivenE2/Deno1
> cat("P(E2|A) = ", PE2givenA, fill = T)
P(E2|A) = 0.3398058
>
> PE3givenA<-PE3*PAgivenE3/Deno1
> cat("P(E3|A) = ", PE3givenA, fill = T)
P(E3|A) = 0.3365696
>
```

■

3.6 RANDOM VARIABLES

A random variable is a function that associates a real number with each element in the sample space. In other words, a random variable (RV) X is a variable that enumerates the outcome of an experiment. Consider a trial of throwing two coins together. The sample space S is {HH, HT, TH, TT}, where T denotes getting a tail and H denotes getting a head. If X is a random variable denoting the number of heads in this experiment, then the value of X can be shown as

$$\text{Outcome:} \quad \text{HH} \quad \text{HT} \quad \text{TH} \quad \text{TT}$$
$$\text{Value of } X: \quad 2 \quad 1 \quad 1 \quad 0$$

Definition of a Random Variable

Consider a probability space, the triplet (S, B, P), where S is the sample space, space of outcomes; B is the σ-field of subsets in S; and P is a probability function of B.

A random variable is a function $X(\omega)$ with domain S and range $(-\infty, \infty)$ such that for every real number a, the event $\{\omega : X(\omega) \leq a\} \in B$.

Random variables are represented by capital letters such as X, Y, and Z. An outcome of the experiment is denoted by $w \cdot X(\omega)$ represents the real number that the random variable X associates with the outcome w. The values, which X, Y, and Z can assume, are represented by lowercase letters, x, y, and z.

If x is a real number, the set of all w in S such that $X(\omega) = x$ is represented by $X = x$. It is denoted as follows:

$$P(X = x) = P\{\omega : X(\omega) = x\}.$$

■ Example 1

Let a gene be investigated to know whether or not it has been expressed. Then,

$$S = \{\omega_1, \omega_2\} \text{ where } \omega_1 = E, \omega_2 = NE,$$

Here $X(\omega)$ takes only two values—a random variable that takes

$$X(\omega) = \begin{cases} 1, & \text{if} \quad \omega = E \\ 0, & \text{if} \quad \omega = NE. \end{cases}$$

■

Distribution Function

Define a random variable X on the space (S, B, P). Then the function

$$F_x(x) = P(X \leq x) = P\{\omega : X(\omega) = x\}, -\infty < x < \infty$$

is called the distribution function (DF) of X.

Sometimes it is convenient to write $F(x)$ instead of $F_x(x)$.

Properties of Distribution Functions
Property 1

F is the distribution function of a random variable X and if $a < b$, then

$$P(a < X \leq b) = F(b) - F(a)$$

Property 2
If F is the distribution function of a random variable X, then

(i) $0 \leq F(x) \leq 1$.

(ii) $F(x) \leq F(y)$ if $x < y$.

3.6.1 Discrete Random Variable

A random variable X can take any value on a real line, but if it takes at most a countable number of values, then the random variable is called a discrete random variable. If a random variable is defined on a discrete sample space, then it is also known as a discrete random variable.

The probability distribution function (PDF) of a discrete random variable is called *a probability mass function*.

Probability Mass Function
A discrete random variable takes at most a countable infinite number of values $x_1, x_2, \ldots$, with each possible outcome x_i. The probability mass function, denoted by $p(x_i); i = 1, 2, \ldots$, denotes the probability of getting the value x_i. It satisfies the following conditions:

(i) $p(x_i) \geq 0 \ \forall i$,

(ii) $\sum_{i=1}^{\infty} p(x_i) = 1$.

■ Example 2
In a microarray experiment, a gene is investigated whether or not it has been expressed. Then

$$S = \{\omega_1, \omega_2\}, \text{ where } \omega_1 = E, \omega_2 = NE$$

$$X(\omega) = \begin{cases} 1, & \text{if } \omega = E \\ 0, & \text{if } \omega = NE \end{cases}.$$

Here $X(\omega)$ takes only two values. Using the previous knowledge about the gene, we know that the gene will express with probability $1/2$. Then, the probability function of the random variable X is given by

$$P(X = 1) = P(\{E\}) = 1/2,$$
$$P(X = 0) = P(\{NE\}) = 1/2.$$ ■

Discrete Distribution Function
If $p(x_i); i = 1, 2, \ldots$, is the probability mass function of a discrete random variable X, then the discrete distribution function of X is defined by

$$F(x) = \sum_{(i:x_i < x)} p(x_i).$$

Also, $p(x_i)$ can be obtained from $F(x)$ as follows:

$$P(x_i) = P(X = x_i) = F(x_i) - F(x_{i-1}).$$

3.6.2 Continuous Random Variable

A continuous random variable X takes all possible values between two specified limits. In other words, a random variable is said to be continuous when its different values cannot be put in 1-1 correspondence with a set of positive integers.

Continuous Probability Functions

In a discrete probability function, sample space has either a finite or a countably infinite number of points. In contrast, in a continuous probability function, the sample space has an uncountably infinite number of points, which are spread over a real line or over some interval of the real line.

A real-valued function defined on a sample space with an uncountable number of outcomes is said to be a continuous probability function if

    a. $f_x(x) \geq 0$,

    b. $\int_s f_x(x)dx = 1$.

Continuous Distribution Function

A continuous distribution function is defined as

$$F_x(x) = P(X \leq x) = \int_{-\infty}^{\infty} f_x(t)dt, \; -\infty < x < \infty$$

where $f_x(x)$ is a continuous density function of a continuous random variable X. Sometimes $F_x(x)$ is written as $F(x)$.

Properties of a Continuous Distribution Function

    1. $0 \leq F(x) \leq 1, -\infty < x < \infty$.

    2. $\frac{d}{dx}F(x) = f(x) \geq 0$.

    3. $F(-\infty) = \lim\limits_{x->-\infty} F(x) = \lim\limits_{x->-\infty} \int_{-\infty}^{x} f(x) = \int_{-\infty}^{-\infty} f(x)dx = 0$.

        $F(\infty) = \lim\limits_{x->\infty} F(x) = \lim\limits_{x->\infty} \int_{\infty}^{x} f(x) = \int_{-\infty}^{\infty} f(x)dx = 1$.

    4. $P(a \leq X \leq b) = \int_a^b f(x)dx = \int_{-\infty}^{b} f(x)dx - \int_{-\infty}^{a} f(x)dx$
            $= P(X \leq b) - P(X \leq a) = F(a) - F(b)$.

3.7 MEASURES OF CHARACTERISTICS OF A CONTINUOUS PROBABILITY DISTRIBUTION

Let $f(x)$ be the PDF of a random variable X where $X_{\varepsilon}(a, b)$. Then

(i) Arithmetic mean $= \int_a^b x(f(x))dx$.

(ii) Harmonic mean: $H = \int_a^b \left(\frac{1}{x}\right)(f(x))dx$.

(iii) Geometric mean: $\log G = \int_a^b \log x(f(x))dx$.

(iv) μ_r(about the origin) $= \int_a^b x^r(f(x))dx$.

μ_r(about the point $x = A$) $= \int_a^b (x - A)^r(f(x))dx$.

μ_r(about the mean) $= \int_a^b (x - mean)^r(f(x))dx$.

(v) Median: If M is the median, then

$$\int_a^M (f(x)dx = \int_M^b (f(x))dx = 1/2.$$

3.8 MATHEMATICAL EXPECTATION

The probability density functions provide all the information contained in the population. However, sometimes it is necessary to summarize the information in single numbers. For example, we can explore the central tendency of the given data by studying the probability density function, but an average will describe the same property in a single number, which allows simple comparison of the central tendency of different populations. In addition, we hope to obtain a number that would inform about the expected value of an observation of the random variable. The most frequently used measure of the central tendency is the expected value.

Definition *Let X be a discrete random variable; then its mathematical expectation is defined as*

$$E(X) = \begin{cases} \int_{-\infty}^{\infty} x f(x)\, dx & \text{if X is continuous,} \\ \sum_{x \in S} x f(x) = \sum_{x \in S} x P(X = x) & \text{if X is discrete,} \end{cases}$$

provided that the integral or sum exists.

If g(x) is a function of X, then

$$E(g(X)) = \begin{cases} \int_{-\infty}^{\infty} g(x)f(x)\, dx & \text{if X is continuous,} \\ \sum_{x \in S} g(x)f(x) = \sum_{x \in S} x P(X = x) & \text{if X is discrete,} \end{cases}$$

provided that the integral or sum exists.

■ Example 1

Let the growth rate per minute (per thousand) of the bacteria "T n T" be given by the PDF $f(x)$ where

$$f(x) = \frac{1}{5}e^{-\frac{x}{5}}, 0 \leq x < \infty.$$

Find the expected growth rate per minute of the bacteria "T n T."

Solution

$$E(X) = \int_0^\infty x\frac{1}{5}e^{-\frac{x}{5}}dx = -x\frac{1}{5} \cdot 5 \cdot e^{-\frac{x}{5}}\Big|_0^\infty + \int_0^\infty e^{-\frac{x}{5}}dx$$

$$= \int_0^\infty e^{\frac{x}{5}}dx = 5.$$

Following is the R-code for example 1:

```
> # Example 1
> # EX = Expected Value
> integrand<-function(x){(x/5)*exp(-x/5)}
> EX<-integrate(integrand, lower = 0, upper = Inf)
> print("EX = ")
[1] "EX = "
> EX
5 with absolute error < 3.8e-05
```

■

■ Example 2

Find the expected value of a number on a die when the die is thrown only one time.

Solution

Let X denote the number on the die. Each number on the die can appear with equal probability $1/6$.

Thus, the expected value of a number obtained on the die is given by

$$E(X) = \frac{1}{6} \times 1 + \frac{1}{6} \times 2 + \frac{1}{6} \times 3 + \frac{1}{6} \times 4 + \cdots + \frac{1}{6} \times 6 = \frac{7}{2}.$$

Following is the R-code for example 2:

```
# Example 2
> # EX = Expected Value, p[i] = Probability getting a number i
```

```
> rm(list = ls())
> for (i in 1:6){p<-1/6}
> p
[1] 0.1666667
> for (i in 1:6){p[i]<-1/6}
> p
[1] 0.1666667 0.1666667 0.1666667 0.1666667 0.1666667 0.1666667
> for (i in 1:6){ex<-i*p[i]}
> ex
[1] 1
> for (i in 1:6){ex[i]<-i*p[i]}
> ex
[1] 0.1666667 0.3333333 0.5000000 0.6666667 0.8333333 1.0000000
> ex1<-sum (ex)
> cat("E(X) = ", ex1, fill = T)
E(X) = 3.5
>
```

■

■ Example 3

Find the expected value of the sum of numbers obtained when a die is thrown twice.

Solution

Let X be the random variable representing the sum of numbers obtained when a die is thrown two times. The probability function X is given by

X	2	3	4	5	6	7
$P(X = x)$	1/36	2/36	3/36	4/36	5/36	6/36

X	8	9	10	11	12
$P(X = x)$	5/36	4/36	3/36	2/36	1/36

$$E(X) = \sum_i p_i x_i$$

$$= \frac{1}{36} \times 2 + \frac{2}{36} \times 3 + \frac{3}{36} \times 4 + \frac{4}{36} \times 5 + \frac{5}{36} \times 6$$

$$+ \frac{6}{36} \times 7 + \cdots + \frac{1}{36} \times 12 = \frac{1}{36} \times 252 = 7.$$

■

Following is the R-code for example 3:

```
> # Example 3
> # EX = Expected  Value, p[i] = Probability getting a number i
> rm(list = ls())
> x<- c(2,3,4,5,6,7,8,9,10,11,12)
> x
[1] 2  3  4  5  6  7  8  9  10  11  12
> for (i in 1:6){p<-1/6}
> p
[1]0.1666667
> for (i in 1:6) {p[i]<-i/36}
> p
[1] 0.02777778 0.05555556 0.08333333 0.11111111 0.13888889 0.16666667
> for (i in 7:11) {p[i]<-(12-i)/36}
> p
[1] 0.02777778 0.05555556 0.08333333 0.11111111 0.13888889 0.16666667
[7] 0.13888889 0.11111111 0.08333333 0.05555556 0.02777778
> for (i in 1:11){ex<-x*p[i]}
> ex
[1] 0.05555556 0.08333333 0.11111111 0.13888889 0.16666667 0.19444444
[7] 0.22222222 0.25000000 0.27777778 0.30555556 0.33333333
> for (i in 1:11){ex[i]<-x[i]*p[i]}
> x
[1] 2 3 4 5 6 7 8 9 10 11 12
> p
[1] 0.02777778 0.05555556 0.08333333 0.11111111 0.13888889 0.16666667
[7] 0.13888889 0.11111111 0.08333333 0.05555556 0.02777778
> ex
[1] 0.05555556 0.16666667 0.33333333 0.55555556 0.83333333 1.16666667
[7] 1.11111111 1.00000000 0.83333333 0.61111111 0.33333333
> ex1<-sum(ex)
> cat("E(X) = ", ex1, fill = T)
E(X) = 7
>
```

3.8.1 Properties of Mathematical Expectation

1. Addition of Expectation

If $X, Y, Z, \ldots, T$ are n random variables, then

$$E(X + Y + Z + \cdots + T) = E(X) + E(Y) + E(Z) + \cdots + E(T),$$

provided all the expectations on the right exist.

2. **Multiplication of Expectation**

   If $X, Y, Z, \ldots, T$ are n independent random variables, then

$$E(XYZ \ldots T) = E(X)E(Y)E(Z) \ldots E(T).$$

3. If X is a random variable and a and b are constants, then

$$E(aX + b) = aE(X) + b,$$

   provided all the expectations exist.

4. Let $g_1(x)$ and $g_2(x)$ be any function of a random variable X. If $a, b,$ and c are constants, then

   a. $E(ag_1(x) + bg_2(x) + c) = aE(g_1(x)) + bE(g_2(x)) + c.$

   b. If $g_1(x) \geq 0$ for all x, then $E(g_1(x)) \geq 0.$

   c. If $g_1(x) \geq g_2(x)$ for all x, then $E(g_1(x)) \geq E(g_2(x)).$

Covariance

If X and Y are two random variables, then covariance between them is defined as

$$\text{Cov}(X, Y) = E[\{X - E(X)\}\{Y - E(Y)\}],$$

or

$$\text{Cov}(X, Y) = E(XY) - E(X)E(Y).$$

If X and Y are independent, then $E(XY) = E(X)E(Y)$ and

$$\text{Cov}(X, Y) = E(X)E(Y) - E(X)E(Y) = 0.$$

Properties of Covariance

1. $\text{Cov}(aX, bY) = ab\text{Cov}(X, Y).$

2. $\text{Cov}(X + a, Y + b) = \text{Cov}(X, Y).$

3. $\text{Cov}\left(\dfrac{X - \overline{X}}{\sigma_X}, \dfrac{Y - \overline{Y}}{\sigma_Y}\right) = \dfrac{1}{\sigma_X \sigma_Y}\text{Cov}(X, Y).$

Variance

Variance of a random variable X is defined as

$$\text{Var}(X) = V(X) = \sigma^2 = E(X^2) - (E(X))^2.$$

Properties of Variance

1. If X is a random variable, then

$$\text{Var}(aX + b) = a^2 \, \text{Var}(X),$$

where a and b are constants.

2. If $X_1, X_2, \ldots X_n$. are n random variables, then

$$\text{Var}\left[\sum_{i=1}^{n} a_i X_i\right] = \sum a_i^2 \, \text{Var}(X_i) + 2 \sum_{i=1}^{n} \sum_{j=1}^{n} a_i a_j \, \text{Cov}(X_i, X_j), \quad i < j,$$

where $a_i = 1; i = 1, 2, \ldots, n$ is a constant.

  a. If $X_1, X_2, \ldots X_n$ are independent, then $\text{Cov}(X_i, X_j) = 0$, and there-fore,

$$\text{Var}(a_1 X_1 + a_2 X_2 + \cdots + a_n X_n)$$
$$= a_1^2 \, \text{Var}(X_1) + a_2^2 \, \text{Var}(X_2) + \cdots + a_n^2 \, \text{Var}(X_n)$$

In particular, if $a_i = 1$, then

$$\text{Var}(X_1 + X_2 + \cdots + X_n)$$
$$= \text{Var}(X_1) + \text{Var}(X_2) + \cdots + \text{Var}(X_n).$$

  b. $\text{Var}(X_1 \pm X_2) = \text{Var}(X_1) + \text{Var}(X_2) \pm 2\text{Cov}(X_1, X_2).$

3. $\text{Var}(X) = 0$ if and only if X is degenerate.

3.9 BIVARIATE RANDOM VARIABLE

One of the major objectives of statistical analysis of data is to establish relation-ships that make it possible to predict one or more variables in terms of others. For example, studies are made to predict the expression level of a gene in terms of the amount of mRNA present, the energy required to break a hydrogen bond in a liquid in terms of its molecular weight, and so forth.

To study two variables, called bivariate random variables, we need the joint distribution of these two random variables.

3.9.1 Joint Distribution

The joint distribution of two random variables X and Y under discrete and continuous cases is defined as follows.

Discrete Case

Two discrete random variables X and Y are said to be jointly distributed if they are defined on the same probability space. The joint probability distribution, p_{ij}, is defined by

$$p_{ij} = P(X = x_i \cap Y = y_i) = p(x_i, y_j), \quad i = 1, 2, \ldots, m; j = 1, 2, \ldots, n,$$

such that

$$\sum_{i=1}^{m} \sum_{j=1}^{n} p(x_i, y_i) = 1.$$

The probability distribution of X, called marginal distribution of X, is defined as

$$p(x_i) = P(X = x_i) = \sum_{j=1}^{n} p_{ij}.$$

Similarly, the marginal probability distribution of Y is defined as

$$p(y_j) = P(Y = y_j) = \sum_{i=1}^{m} p_{ij}.$$

Also, the conditional probability function of X given Y is defined as

$$p\left(X = x_i | Y = y_j\right) = \frac{p(x_i, y_j)}{p(y_j)}.$$

Similarly, the conditional probability function of Y given X is given by

$$p\left(Y = y_j | X = x_i\right) = \frac{p(x_i, y_j)}{p(x_i)}.$$

Note, also, that the two random variables X and Y will be independent if $p(x_i, y_j) = p(x_i)p(y_j)$.

Continuous Case

If X and Y are continuous random variables, then their joint probability function is given by $f_{X,Y}(x, y)$ provided that

$$f_{X,Y}(x, y) = \int_{-\infty}^{x} \int_{-\infty}^{y} f_{XY}(x, y)dxdy,$$

such that

$$\int_{-\infty}^{\infty} \int_{-\infty}^{\infty} f_{XY}(x, y)\,dx\,dy = 1.$$

The marginal distribution of the random variable X is given by

$$g_x(x) = \int_{-\infty}^{\infty} f_{X,Y}(x, y)\,dy.$$

The marginal distribution of the random variable Y is given by

$$h_Y(y) = \int_{-\infty}^{\infty} f_{X,Y}(x, y)\,dx.$$

The conditional distribution of the random variable Y given X is defined by

$$f_{Y|X}(Y|X) = \frac{f_{XY}(x, y)}{g_x(x)}.$$

The conditional distribution of the random variable X given Y is defined by

$$f_{X|Y}(X|Y) = \frac{f_{XY}(x, y)}{h_Y(y)}.$$

Two random variables X and Y (discrete or continuous) with joint PDFs $f_{X|Y}(x|y)$ and marginal PDFs $g_x(x)$ and $h_Y(y)$, respectively, are said to be independent if

$$f_{XY}(x, y) = g_X(x)h_Y(y).$$

■ Example 1

A gene is studied to see whether the change in temperature, denoted by a random variable X, and monochromatic light of different wavelengths, denoted by a random variable Y, lead to a change in expression level of the gene. The following table gives the bivariate probability distribution of X and Y. Find

   a. $P(X \leq 2, Y = 2)$.

   b. $P(X \leq 1)$.

   c. $P(Y = 2)$.

   d. Marginal distributions of X and Y.

e. Conditional distribution of $X = 1$ given $Y = 2$.

Y \ X	0	1	2	3	4	5	6
0	0	0	0	$\frac{1}{32}$	$\frac{1}{32}$	$\frac{3}{32}$	$\frac{3}{32}$
1	$\frac{1}{16}$	$\frac{1}{16}$	$\frac{1}{16}$	$\frac{1}{8}$	$\frac{2}{8}$	0	$\frac{1}{8}$
2	$\frac{1}{64}$	$\frac{1}{64}$	$\frac{1}{64}$	$\frac{1}{64}$	0	0	0

Solution

We first find the marginal probabilities of X and Y by using

$$p(x_i) = \sum_{j=1}^{n} p_{ij} = \sum_{j=1}^{n} P(x = x_i, Y = y_j),$$

which is obtained by adding entries in the respective column, and

$$p(y_j) = \sum_{i=1}^{m} p_{ij} = \sum_{i=1}^{m} p(X = x_i, Y = y_j),$$

which is obtained by adding entries in the respective row.

Y \ X	0	1	2	3	4	5	6	$p(y)$
0	0	0	0	$\frac{1}{32}$	$\frac{1}{32}$	$\frac{3}{32}$	$\frac{3}{32}$	$\frac{8}{32}$
1	$\frac{1}{16}$	$\frac{1}{16}$	$\frac{1}{16}$	$\frac{1}{8}$	$\frac{2}{8}$	0	$\frac{1}{8}$	$\frac{11}{16}$
2	$\frac{1}{64}$	$\frac{1}{64}$	$\frac{1}{64}$	$\frac{1}{64}$	0	0	0	$\frac{4}{64}$
$p(x)$	$\frac{5}{64}$	$\frac{5}{64}$	$\frac{5}{64}$	$\frac{11}{64}$	$\frac{9}{32}$	$\frac{3}{32}$	$\frac{7}{32}$	$\Sigma p(x) = 1$ $\Sigma p(y) = 1$

a. $P(X \leq 2, Y = 2)$

$$= P(X = 0, Y = 2) + P(X = 1, Y = 2) + P(X = 2, Y = 2)$$

$$= \frac{1}{64} + \frac{1}{64} + \frac{1}{64} = \frac{3}{64}.$$

b. $P(X \leq 3) = P(X = 0) + P(X = 1) + P(X = 2) + P(X = 3)$

$$= \frac{5}{64} + \frac{5}{64} + \frac{5}{64} + \frac{11}{64} = \frac{26}{64}.$$

c. $P(Y = 2) = \frac{4}{64}.$

d. The marginal distribution of Y is given by

Y	0	1	2	
$p(y) = P(Y = y)$	$\frac{8}{32}$	$\frac{11}{16}$	$\frac{4}{64}$	$\sum p(y) = 1$

The marginal distribution of X is given by

X	0	1	2	3	4	5
$p(x) = P(X = x)$	$\frac{5}{64}$	$\frac{5}{64}$	$\frac{5}{64}$	$\frac{11}{64}$	$\frac{9}{32}$	$\frac{3}{32}$

e. We have

$$P(X = x | Y = 2) = \frac{P(X = x \cap Y = 2)}{P(Y = 2)}.$$

$$P(X = 1 | Y = 2) = \frac{P(X = 1 \cap Y = 2)}{P(Y = 2)} = \frac{1/64}{4/64} = \frac{1}{4}.$$

Following is the R-code for example 1.

```
> #Example 1
> rm(list = ls())
> genetable<-matrix(c(0,0,0,1/32,1/32,3/32,3/32,1/16,
+1/16,1/16,1/8,2/8,0,1/8,1/64,1/64,1/64,1/64,0,0,0),
+nrow = 3, byrow = T)
> print("Joint Probability Matrix")
[1] "Joint Probability Matrix"
> genetable
     [,1]    [,2]    [,3]    [,4]    [,5]   [,6]   [,7]
[1,] 0.000000 0.000000 0.000000 0.031250 0.03125 0.09375 0.09375
[2,] 0.062500 0.062500 0.062500 0.125000 0.25000 0.00000 0.12500
[3,] 0.015625 0.015625 0.015625 0.015625 0.00000 0.00000 0.00000
> rownames(genetable)<-c("0","1","2")
> colnames(genetable)<-c("0","1","2","3","4","5","6")
> print("Joint Probability Matrix with Headings")
[1] "Joint Probability Matrix with Headings"
> genetable
  0 1 2 3 4 5 6
0 0.000000 0.000000 0.000000 0.031250 0.03125 0.09375 0.09375
1 0.062500 0.062500 0.062500 0.125000 0.25000 0.00000 0.12500
2 0.015625 0.015625 0.015625 0.015625 0.00000 0.00000 0.00000
>
> #Checking whether sum of all probabilities is 1 or not
> sum(genetable)
[1] 1
>
> py1<-sum(genetable[1,])
```

```
> py2<-sum(genetable[2,])
> py3<-sum(genetable[3,])
> print("Marginal Probabilities P[Y = y]")
[1] "Marginal Probabilities P[Y = y]"
> cat("P(Y = 0) = ", py1, fill = T)
P(Y = 0) = 0.25
> cat("P(Y = 1) = ", py2, fill = T)
P(Y = 1) = 0.6875
> cat("P(Y = 2) = ", py3, fill = T)
P(Y = 2) = 0.0625
>
> px1<-sum(genetable[,1])
> px2<-sum(genetable[,2])
> px3<-sum(genetable[,3])
> px4<-sum(genetable[,4])
> px5<-sum(genetable[,5])
> px6<-sum(genetable[,6])
> px7<-sum(genetable[,7])
> print("Marginal Probabilities P[X = x]")
[1] "Marginal Probabilities P[X = x]"
> cat("P(X = 0) = ", px1, fill = T)
P(X = 0) = 0.078125
> cat("P(X = 1) = ", px2, fill = T)
P(X = 1) = 0.078125
> cat("P(X = 2) = ", px3, fill = T)
P(X = 2) = 0.078125
> cat("P(X = 3) = ", px4, fill = T)
P(X = 3) = 0.171875
> cat("P(X = 4) = ", px5, fill = T)
P(X = 4) = 0.28125
> cat("P(X = 5) = ", px6, fill = T)
P(X = 5) = 0.09375
> cat("P(X = 6) = ", px7, fill = T)
P(X = 6) = 0.21875
>
> #converting P[X = x] in a matrix form
> mpx<-matrix(c(px1,px2,px3,px4,px5,px6,px7),nrow = 1)
>
> #Combining P[X = x, Y = y] and P[X = x]
> ppx<-rbind(genetable,mpx)
>
> #Converting P[Y = y] in a matrix form with last row as sum of P[X = x]
> mpy<-matrix(c(py1,py2,py3,1), nrow = 4, byrow = T)
>
> #Combining P[X = x, Y = y], P[X = x] and P[Y = y]
```

```
> ppxy<-cbind(ppx,mpy)
>
> #Inserting row and column headings
> rownames(ppxy)<-c("0","1","2","P[X = x}")
> colnames(ppxy)<-c("0","1","2","3","4","5","6","P[Y = y]")
>
> print("Joint Probability Matrix with marginal Probabilities)"
[1] "Joint  Probability Matrix with marginal Probabilities"
> ppxy
      0  1  2  3  4  5   6 P[Y = y]
0   0.000000 0.000000 0.000000 0.031250 0.03125 0.09375 0.09375 0.2500
1   0.062500 0.062500 0.062500 0.125000 0.25000 0.00000 0.12500 0.6875
2   0.015625 0.015625 0.015625 0.015625 0.00000 0.00000 0.00000 0.0625
P[X = x] 0.078125 0.078125 0.078125 0.171875 0.28125 0.09375 0.21875 1.0000
>
> #Solution Part a
> # pl2y2 = P[x< = 2,y = 2]
> # pr1 = probabilities
> pr1<-c(ppxy[3,1],ppxy[3,2],ppxy[3,3])
> cat("P(X = 0, Y = 2) = ", ppxy[3,1], fill = T)
P(X = 0, Y = 2) = 0.01 5625
> cat("P(X = 1, Y = 2) = ", ppxy[3,2], fill = T)
P(X = 1, Y = 2) = 0.015625
> cat("P(X = 2, Y = 2) = ", ppxy[3,3], fill = T)
P(X = 2, Y = 2) = 0.015625
> pl2y2<-sum(pr1)
> cat("P(X< = 2, Y = 2) = ", pl2y2, fill = T)
P(X< = 2, Y = 2) = 0.046875
> #Solution  Part b
> #pxle3 = P[X< = 3]
> #pr2 = probabilities
> pr2<-c(mpx[1],mpx[2],mpx[3],mpx[4])
> pxle3<- sum(pr2)
> cat("P[X< = 3] = ", pxle3, fill = T)
P[X< = 3] = 0.40625
>
> #Solution Part C
> #pr3 = P[Y = 2]
> pr3<-py3
> cat("P[Y = 3] = ", pr3, fill = T)
P[Y = 3] = 0.0625
>
> #Solution Part D
> rownames(mpy)<-c("P[Y = 0]","P[Y = 1]","P[Y = 2","Sum P[Y = y]")
> print("Marginal Distribution of Y, P[Y = y], and Sum of P[Y = y]")
```

[1] "Marginal Distribution of Y, P[Y = y], and Sum of P[Y = y]"
> mpy
 [,1]

P[Y = 0] 0.2500
P[Y = 1] 0.6875
P[Y = 2] 0.0625
Sum P[Y =y] 1.0000
>
> colnames(mpx)<-c("P[X = 0]","P[X = 1]","P[X = 2]","P[X = 3]","P[X = 4]","P[X = 5]","P[X = 6]")
> print("Marginal Distribution of X, P[X = x]")
[1] "Marginal Distribution of X, P[X = x]"
> mpx
P[X = 0] P[X = 1] P[X = 2] P[X = 3] P[X = 4] P[X = 5] P[X = 6]
[1,] 0.078125 0.078125 0.078125 0.171875 0.28125 0.09375 0.21875
>
> #Solution Part E
>
> #px1gy2 = P[X = 1|Y = 2)
> px1gy2<-genetable[3,2]/py3
> cat("P[X = 1|Y = 2] = ",px1gy2,fill = T)
P[X = 1|Y = 2] = 0.25
>   ■

■ Example 2

In whole intact potato (*Solanum tuberosum* L.) plants, the gene families of class-I patatin and proteinase inhibitor II (Pin 2) are expressed in the tubers. A study was conducted to see the effect of room pressure on the gene expression. The joint probability density function of expression (X) and pressure (Y) is given by

$$f(x, y) = \begin{cases} 4xy, & 0 \le x \le 1, \text{ and } 0 \le y \le 1, \\ 0, & \text{otherwise.} \end{cases}$$

    (i) Find the marginal density functions of X and Y.

    (ii) Find the conditional density functions of X and Y.

    (iii) Check for independence of X and Y.

Solution

We find that $f(x, y) \ge 0$ and

$$\int_0^1 \int_0^1 4xy\,dx\,dy = 2 \int_0^1 x\,dx = 1.$$

Thus, $f(x, y)$ represents a joint probability density function.

(i) The marginal probability density functions of X and Y are given by

$$g_X(x) = \int_{-\infty}^{\infty} f_{XY}(x, y)dy = \int_{0}^{1} 4xy \, dy = 2x, \quad 0 < x < 1;$$

$$h_Y(y) = \int_{-\infty}^{\infty} f_{XY}(x, y)dy = \int_{0}^{1} 4xy \, dx = 2y, \quad 0 < y < 1.$$

(ii) The conditional density function of Y given X is

$$f_{Y/X}(y|x) = \frac{f_{XY}(x, y)}{g_X(x)} = \frac{4xy}{2x} = 2y, \quad 0 < x < 1; 0 < y < 1.$$

The conditional density function of X given Y is

$$f_{X/Y}(x|y) = \frac{f_{XY}(x, y)}{h_Y(y)} = \frac{4xy}{2y} = 2x, \quad 0 < x < 1; 0 < y < 1.$$

(iii) We find that

$$g_X(x)h_Y(y) = 2x \cdot 2y = 4xy = f_{XY}(x, y).$$

Hence, X and Y are independent.

Following is the R-code for example 1.

```
#Example 1
#checking whether f(x,y) is a probability function
Pxy<-integrate (function(y){
sapply(y,function(y){
integrate(function(x){
sapply(x,function(x)(4*x*y))
}, 0, 1)$value
})
}, 0,1)
print("Value of int int f(x,y)dxdy")
Pxy
#Part A
# Assume x to be a constant
# Px1 = f(x)
Px1<-integrate(function(y){
4*y
}, 0,  1)$value
```

```
Px1
# the pdf f(x) = x*Px1
# Assume y  to be a constant
# Py1 = f(y)
Py1<-integrate(function(x){
4*x
}, 0,  1)$value
Py1
# the pdf f(y) = y*Py1
```
■

3.10 REGRESSION

The term *regression* was used by Francis Galton to indicate the relationships in the theory of hereditary. Formally, the term *regression* means determining the conditional expectation $\mu_{Y|X}$, that is the average value of Y for the given value of X when the joint distribution of two random variables X and Y is given. For example, one would like to relate the incidence of cancer to its many contributing causes such as diet, heredity, pollution, and smoking.

If $f(x, y)$ is the joint density function of two random variables X and Y, then the bivariate regression equation of Y on X is given by

$$\mu_{Y|X} = E(Y|X) = \int_{-\infty}^{\infty} yf(y|x)dy.$$

The bivariate regression equation of X on Y is given by

$$\mu_{X|Y} = E(X|Y) = \int_{-\infty}^{\infty} xf(x|y)dx.$$

In the discrete case, the probability densities are replaced by probability mass function, and the integrals are replaced by sums.

■ Example 1

The amount of clean wool produced by individual sheep is important in a breeding program designed for wool improvement because clean wool is the most accurate measure of an animal's wool-producing ability (Sidwell, 1956). A sample is taken from a sheep population, and clean grease fleece weights are measured. The joint density function, $f(x, y)$, gives the average clean wool obtained from both parents (midparents' average) denoted by X, and their offspring, denoted by Y. The density function is given by

$$f(x, y) = \begin{cases} xe^{-(x+y)}, & \text{for } x > 0, \text{ and } y > 0, \\ 0, & \text{elsewhere.} \end{cases}$$

Find the regression equation of offspring on midparents.

Solution

We find that the marginal density function of X is given by

$$g(x) = \int_0^\infty f(x, y)dy = \int_0^\infty xe^{-(x+y)}dy = xe^{-x}$$

Thus,

$$g(x) = \begin{cases} xe^{-x}, & \text{for } x > 0, \\ 0, & \text{elsewhere.} \end{cases}$$

Hence, the conditional density of Y given $X = x$ is given by

$$f(y|x) = \frac{f(x, y)}{g(x)} = \frac{xe^{-(x+y)}}{xe^{-x}} = e^{-y}, \text{ for } y > 0.$$

Thus,

$$\mu_{Y|X} = E(Y|X) = \int_{-\infty}^\infty yf(y|x)dy$$

$$= \int_0^\infty y \cdot e^{-y}dy = 1.$$

∎

3.10.1 Linear Regression

Let the linear regression equation be given by

$$\mu_{Y|X} = \alpha + \beta x,$$

where α and β are constants that are known as regression coefficients. The linear regression is important due to the ease with which mathematical treatments can be applied, and also in the case of bivariate normal distribution, the regression of Y on X or X on Y is in fact linear in nature.

The linear regression equation can be written in terms of variances, covariance, and means as follows.

Let $E(X) = \mu_1, E(Y) = \mu_1, \text{Var}(X) = \sigma_1^2, \text{Var}(Y) = \sigma_2^2$ and $\text{Cov}(X, Y) = \sigma_{12}$. Also, the correlation coefficient is defined as

$$\rho = \frac{\sigma_{12}}{\sigma_1 \sigma_2}.$$

If the regression of Y on X is linear, then

$$\alpha = \mu_2 - \rho \frac{\sigma_2}{\sigma_1} \mu_1$$

and

$$\beta = \rho \frac{\sigma_2}{\sigma_1}.$$

Thus, we can write the linear regression equation of Y on X as

$$\mu_{Y|X} = \mu_2 + \rho \frac{\sigma_2}{\sigma_1} (X - \mu_1).$$

Similarly, the regression of X on Y is given by

$$\mu_{X|Y} = \mu_1 + \rho \frac{\sigma_1}{\sigma_2} (Y - \mu_2).$$

If $\rho = 0$, then $\sigma_{12} = 0$, and therefore $\mu_{Y|X}$ does not depend on x (or $\mu_{X|Y}$ does not depend on y), which makes two random variables X and Y uncorrelated. Thus, if two random variables are independent, they are also uncorrelated, but the reverse is not always true.

3.10.2 The Method of Least Squares

Most times, we are given a set of data, but the joint distribution of the random variables under consideration is not known. If an aim is to estimate α, and β, then the method of least squares is one of the most commonly used methods in this situation. Let there be n points, namely, $(x_1, y_1), (x_1, y_1), \ldots, (x_n, y_n)$. Let $p(x)$ be the polynomial that is closest to the given points and has the form

$$p(x) = \sum_{k=0}^{m} \alpha_k x^k,$$

where $\alpha_0, \alpha_1, \ldots, \alpha_m$ are to be determined. The method of least squares selects those α_ks as solutions, which minimizes the sum of squares of the vertical distances from the data points to the polynomial $p(x)$. Thus, the polynomial $p(x)$

will be called best if its coefficient minimizes the sum of squares of errors, q, where

$$q = \sum_{i=1}^{n} e_i^2 = \sum_{i=1}^{n} [y_i - p(x_i)]^2.$$

For n given points $(x_1, y_1), (x_1, y_1), \ldots, (x_n, y_n)$, the straight line $y = \alpha + \beta x$ minimizes

$$q = \sum_{i=1}^{n} e_i^2 = \sum_{i=1}^{n} [y_i - (\alpha + \beta x_i)]^2.$$

Solving equation q, we find that the least square estimates of regression coefficients are the values $\hat{\alpha}$ and $\hat{\beta}$, for which q is a minimum. After differentiating q with respect to $\hat{\alpha}$ and $\hat{\beta}$ and equating the system of equation to zero, we determine that

$$\hat{\beta} = \frac{n \left(\sum_{i=1}^{n} x_i y_i \right) - \left(\sum_{i=1}^{n} x_i \right) \left(\sum_{i=1}^{n} y_i \right)}{n \left(\sum_{i=1}^{n} x_i^2 \right) - \left(\sum_{i=1}^{n} x_i \right)^2},$$

and

$$\hat{\alpha} = \frac{\sum_{i=1}^{n} y_i - \hat{\beta} \cdot \sum_{i=1}^{n} x_i}{n}.$$

To simplify the formulas, we use the following notations:

$$S_{xx} = \sum_{i=1}^{n} (x_i - \bar{x})^2 = \sum_{i=1}^{n} x_i^2 - \frac{1}{n} \left(\sum_{i=1}^{n} x_i \right)^2,$$

$$S_{yy} = \sum_{i=1}^{n} (y_i - \bar{y})^2 = \sum_{i=1}^{n} y_i^2 - \frac{1}{n} \left(\sum_{i=1}^{n} y_i \right)^2,$$

$$S_{xy} = \sum_{i=1}^{n} (x_i - \bar{x})(y_i - \bar{y}) = \sum_{i=1}^{n} x_i y_i - \frac{1}{n} \left(\sum_{i=1}^{n} x_i \right) \left(\sum_{i=1}^{n} y_i \right).$$

Thus, $\hat{\alpha}$ and $\hat{\beta}$ can be written as

$$\hat{\beta} = \frac{S_{xy}}{S_{xx}},$$

and

$$\hat{\alpha} = \bar{y} - \hat{\beta} \cdot \bar{x}.$$

■ Example 2

A scientist studied the average number of bristles (Y) on a segment of *Drosophila melanogaster* in five female offspring and number of bristles in the mother (dam) (X) of each set of five daughters.

The following data set was collected:

Family	1	2	3	4	5	6	7	8	9	10
X	6	9	6	7	8	7	7	7	9	9
Y	8	6	7	7	8	7	7	9	9	9

    a. Find the regression equation of daughters on the dam.

    b. Also find the average number of bristles in five female offspring when the number of bristles in the mother (dam) of each of the five daughters is 11.

Solution

    a. We omit the limits of summation in notation for simplicity. We get $n = 10$, $\Sigma x = 75$, $\Sigma y = 77$, $\Sigma x^2 = 575$, $\Sigma y^2 = 603$, $\Sigma xy = 580$. Thus,

$$S_{xx} = 603 - \frac{1}{10}(75)^2 = 40.5,$$

$$S_{xy} = 580 - \frac{1}{10}(75)(77) = 2.5,$$

$$\hat{\beta} = \frac{S_{xy}}{S_{xx}} = \frac{2.5}{40.5} = 0.0617.$$

and

$$\hat{\alpha} = \bar{y} - \hat{\beta} \cdot \bar{x} = 7.7 - 0.0617 \cdot 7.5 = 7.237.$$

Thus, the least square regression equation is

$$\hat{y} = 7.237 + 0.0617x.$$

    b. When $x = 11$,

$$\hat{y} = 7.237 + 0.0617 \cdot 11 = 7.9159 \cong 8.$$

Thus, there are eight bristles on a segment of *Drosophila melanogaster* in five female offspring when there are 11 bristles in the mother (dam) of each set of five daughters.

Following is the R-code for example 1.

```
> #Example 1
> #Regression and Correlation
> rm(list = ls())
> x<-c(6,9,6,7,8,7,7,9,9)
> y<-c(8,6,6,7,8,7,9,9,9)
> # show values of X and Y
> x
[1] 6 9 6 7 8 7 7 7 9 9
> y
[1] 8 6 6 7 8 7 7 9 9 9
>
> #Scatter Plot between X and Y
>
> plot(x,y)
> linreg = lm(x ~ y)
> linreg
Call:
lm(formula = x ~ y)
Coefficients:
(Intercept)          y
      5.0484     0.3226 >
> # fitted line on the scatter plot
> # use command "abline(intercept, coefficient of y)"
> abline(5.0484,0.3226)
> # Confidence intervals and tests
>
> summary(linreg)
Call:
lm(formula = x ~ y)
Residuals:
Min    1Q    Median    3Q    Max
-1.6290 -0.7903 -0.3065 0.8790 2.0161
Coefficients:
Estimate Std. Error t value Pr(>|t|)
(Intercept) 5.0484 2.5821 1.955 0.0863.
y 0.3226 0.3362 0.960 0.3653
---
Signif. codes: 0 '***' 0.001 '**'0.01 '*' 0.05'.' 0.1 "1
Residual standard error: 1.184 on 8 degrees of freedom
Multiple R-squared: 0.1032, Adjusted R-squared: -0.008871
F-statistic: 0.9209 on 1 and 8 DF, p-value: 0.3653
```

```
>
> #Estimated value for given x = 11
> x1<-11
> yest<-5.0484 + 0.3226*x1
>
> #print estimated value and x = 11
>
> x1
[1] 11
> yest
[1] 8.597
> #Predicting all values of Y for X values
> pred = predict(linreg)
> pred
       1        2        3        4        5        6        7        8
7.629032 6.983871 6.983871 7.306452 7.629032 7.306452 7.306452 7.951613
       9       10
7.951613 7.951613
>
```

Scatter plot with fitted line

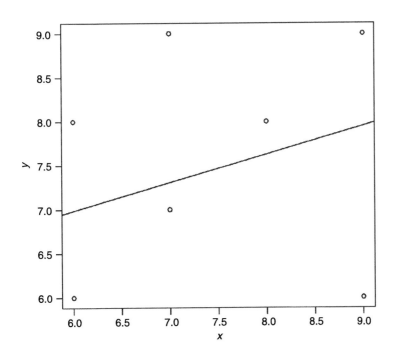

3.11 CORRELATION

Sometimes it is of interest to find out whether there is any correlation between two variables under study. If the change in one variable leads to a change in other variable, then two variables are said to be correlated. If the increase in one variable leads to an increase in the other variable, then two variables are said to be positively correlated; and on the other hand, if the increase (or decrease) in one variable leads to a decrease (or increase) in the other variable, then the two variables are said to be negatively related. For example, an increase in price leads to a decrease in demand; thus, price and demand are negatively related. An increase in income leads to an increase in expenditure; therefore, income and expenditure are positively related.

The correlation coefficient is a measure of the extent of linear relationship between two variables X and Y. The population correlation coefficient ρ is given by

$$\rho = \frac{\text{Cov}(X, Y)}{\sqrt{\text{Var}(X) \cdot \text{Var}(Y)}}.$$

If the population correlation coefficient between X and Y is unknown but we have the values of a random sample $\{(x_i, y_i); i = 1, 2, \ldots, m\}$ from the population, then we can estimate p by sample correlation coefficient, r, given by

$$r = \frac{S_{xy}}{\sqrt{S_{xx}S_{yy}}},$$

where

$$S_{xx} = \sum_{i=1}^{n}(x_i - \bar{x})^2 = \sum_{i=1}^{n}x_i^2 - \frac{1}{n}\left(\sum_{i=1}^{n}x_i\right)^2,$$

$$S_{yy} = \sum_{i=1}^{n}(y_i - \bar{y})^2 = \sum_{i=1}^{n}y_i^2 - \frac{1}{n}\left(\sum_{i=1}^{n}y_i\right)^2,$$

$$S_{xy} = \sum_{i=1}^{n}(x_i - \bar{x})(y_i - \bar{y}) = \sum_{i=1}^{n}x_iy_i - \frac{1}{n}\left(\sum_{i=1}^{n}x_i\right)\left(\sum_{i=1}^{n}y_i\right).$$

Properties of Correlation Coefficient

(1) $|\rho| \leq 1$.

(2) $|\rho| = 1$ if and only if $Y = \alpha + \beta x$ for some constants α and β.

(3) If X and Y are independent variables, then $\rho = 0$.

■ Example 1

A scientist desires to measure the level of coexpression between two genes. The level of coexpression is calculated by finding the correlation between the expression profiles of the genes. If two expression profiles are alike, then the correlation between two genes will be high. The expression levels of two genes X and Y in 10 experiments are given here. Find the sample correlation coefficient between two genes.

Experiment	1	2	3	4	5	6
X	0.512	0.614	0.621	0.519	0.525	0.617
Y	0.423	0.523	0.444	0.421	0.456	0.545
Experiment	7	8	9	10		
X	0.775	0.555	0.674	0.677		
Y	0.672	0.423	0.567	0.576		

Solution

From the given data, we get $n = 10$, $\Sigma x = 6.089$, $\Sigma y = 5.05$, $\Sigma x^2 = 3.771$, $\Sigma y^2 = 2.615$, $\Sigma xy = 3.207$.

Thus,

$$S_{xx} = 3.771 - \frac{1}{10}(6.089)^2 = 0.063,$$

$$S_{yy} = 2.615 - \frac{1}{10}(5.05)^2 = 0.064,$$

$$S_{xy} = 3.135 - \frac{1}{10}(6.089)(5.05) = 0.0603,$$

$$r = \frac{S_{xy}}{\sqrt{S_x S_{yy}}} = \frac{0.0603}{\sqrt{0.063 \cdot 0.064}} = 0.9319.$$

Thus, we find that two genes are highly correlated.

Following is the R-code for example 1.

```
#Example 1
> #Correlation
> rm(list = ls())
> x<-c(0.512,0.614,0.621,0.519,0.525,0.617,0.775,0.555,0.674,0.677)
> y<-c(0.423,0.523,0.444,0.421,0.456,0.545,0.672,0.423,0.567,0.576)
> # show values of X and Y
> x
[1] 0.512 0.614 0.621 0.519 0.525 0.617 0.775 0.555 0.674 0.677
> y
```

[1] 0.423 0.523 0.444 0.421 0.456 0.545 0.672 0.423 0.567 0.576

\>

\> #Scatter Plot between X and Y

\>

\> plot(x,y)

\> #Correlation between X and Y

\> cor(x,y)

[1] 0.9319725

\>

■

3.12 LAW OF LARGE NUMBERS AND CENTRAL LIMIT THEOREM

A sequence of random variables $X_1, X_2, \ldots, X_n$ is said to converge in probability to a constant c if for any $\varepsilon > 0$,

$$\lim_{n \to \infty} P(|X_n - c| < \varepsilon) = 1.$$

Symbolically, it is written as

$$X_n \xrightarrow{p} c \text{ as } n \to \infty.$$

Chebyshev's Theorem

If $X_1, X_2, \ldots, X_n$ is a sequence of random variables and if mean μ_n and standard deviation σ_n of X_n exists for all n, and if $\sigma_n \to 0$ as $n \to \infty$, then

$$X_n \xrightarrow{p} \mu_n \text{ as } n \to \infty.$$

Law of Large Numbers

For any positive constant c, the probability that the sample mean $\overline{X}$ will take on a value between $\mu - c$ and $\mu + c$ is at least $1 - \frac{\sigma^2}{nc^2}$ and the probability that $\overline{X}$ will take on a value between $\mu - c$ and $\mu + c$ approaches 1 as $n \to \infty$.

Central Limit Theorem

If $X_1, X_2, \ldots, X_n$ is a random sample from an infinite population with mean μ and variance σ^2, then the limiting distribution of

$$Z = \frac{\overline{X} - \mu}{\sigma/\sqrt{n}}$$

as $n \to \infty$ is the standard normal distribution.

In other words, the central limit theorem can be stated as follows: If X_i, $(i = 1, 2, \ldots, n)$ is an independent random variable such that $E(X_i) = \mu_i$ and

$V(X_i) = \sigma_i^2$, then under general conditions, the random variables $S_n = X_1 + X_2 + X_3 + \cdots + X_n$ are asymptotically normal with mean μ and standard deviation σ where

$$\mu = \sum_{i=1}^{n} \mu_i; \sigma^2 = \sum_{i=1}^{n} \sigma_i^2.$$

Lindeberg-Levy Theorem

The Lindeberg-Levy Theorem is a case of a central limit theorem for identically distributed variables. If $X_1, X_2, \ldots, X_n$ are independently and identically distributed random variables with $E(X_i) = \mu_1$ and $V(X_i) = \sigma_1^2$ for all $i = 1, 2, \ldots, n$, then the sum $S_n = X_1 + X_2 + X_3 + \cdots + X_n$ is asymptotically normal with mean $\mu = n\mu_1$ and variance $\sigma^2 = n\sigma_1^2$ provided $E(X_i^2)$ exists for all $i = 1, 2, \ldots n$.

Liapounoff's Central Limit Theorem

If $X_1, X_2, \ldots, X_n$ are independently distributed random variables with

$$E(X_i) = \mu_i \text{ and } V(X_i) = \sigma_i^2, i = 1, 2, \ldots, n.$$

Let us suppose that the third absolute moment of X_l about its mean

$$\rho_i^3 = E\{|X_i - \mu_i|^3\}; i = 1, 2 \ldots n$$

is finite. Let $\rho^3 = \sum_{i=1}^{n} \rho_i^3$.

If $\lim_{n \to \infty} \frac{\rho}{\sigma} = 0$, then the sum $S_n = X_1 + X_2 + X_3 + \cdots + X_n$ is asymptotically $N(\mu, \sigma^2)$, where

$$\mu = \sum_{i=1}^{n} \mu_i \text{ and } \sigma^2 = \sum_{i=1}^{n} \sigma_i^2.$$

## CHAPTER 4

Special Distributions, Properties, and Applications

4.1 INTRODUCTION

The probability density function $f_x(x)$ defined over a sample space S needs to satisfy two conditions: (a) $f_x(x) \geq 0$, and (b) $\int_S f_x(x)\, dx = 1$. Thus, theoretically, there can be an infinite number of probability density functions that can satisfy these two conditions but may not be practically useful for the explanation of real-world phenomena. There are many PDFs that model real-world phenomena based on the assumptions on the measurements. It makes sense to investigate the properties of those PDFs that model real-world phenomena in more detail.

4.2 DISCRETE PROBABILITY DISTRIBUTIONS

We will consider some of the important distributions that are associated with discrete random variables. These distributions have a wide range of applications in bioinformatics.

4.3 BERNOULLI DISTRIBUTION

The outcome of a Bernoulli trial is one of two mutually exclusive and exhaustive events, generally called success or failure. Let a random variable X take two values 0 and 1, with probability q and p, respectively; that is, $P(X = 1) = p$, $P(X = 0) = q = 1 - p$. This random variable X is called a Bernoulli variate and is said to have a Bernoulli distribution. The PDF of X can be written as

$$f(x) = p^x(1 - p)^{1-x}, x = 0, 1. \tag{4.3.1}$$

The moment of origin is given by

$$\mu'_k = E(X^k) = \sum_{x=0}^{1} x^k p^x(1 - p)^{1-x} = 0^k \cdot (1 - p) + 1^k \cdot p = p; \quad k = 1, 2, \ldots$$

In particular,

$$\mu'_1 = E(X) = p, \mu'_2 = E(X^2) = p.$$

Thus,

$$\mu_2 = \text{Var}(X) = p^2 - p = p(1 - p) = pq. \tag{4.3.2}$$

In a sequence of Bernoulli trials, generally we are interested in the number successes and not in the order of their occurrence. This leads to a new distribution known as binomial distribution.

4.4 BINOMIAL DISTRIBUTION

Let there be n independent Bernoulli trials, where the probability p of success in any trial is constant for each trial. Then $q = 1 - p$ is the probability of failure in any trial. A sequence of Bernoulli trials with x success, and consequently $(n - x)$ failures in n independent trials, in a specified order can be represented as SSSSFFFSFS....FS (where S represents a success, and F represents a failure).

Thus, the probability of x successes and $(n - x)$ failures is given by

$$P(SSSSFFFSFS....FS) = P(S)P(S)P(S)....P(F)P(S)$$
$$= p \cdot p \cdot p \cdots q \cdot p$$
$$= p^x q^{n-x}. \tag{4.4.1}$$

Since the x success in n trials can occur in $\binom{n}{x}$ ways and the probability for each of these ways is $p^x q^{n-x}$, the probability of x successes in n trials in any order is given by

$$\binom{n}{x} p^x q^{n-x}. \tag{4.4.2}$$

Definition *A random variable X is said to follow binomial distribution if its PDF is given by*

$$P(X = x) = p(x) = \begin{cases} \binom{n}{x} p^x q^{n-x}, & x = 0, 1, 2, \ldots, n, q = 1 - p, \\ 0, & \text{otherwise.} \end{cases} \tag{4.4.3}$$

The two independent constants n and p are known as the parameters of the binomial distribution. The notation Bin(n, p) is used to denote that the random variable X follows the binomial distribution with parameters n and p. There are four defining conditions with the binomial distribution, which follow:

1. *Each trial results in one of the two possible outcomes.*

2. *All trials in an experiment are independent.*

3. *Probability of success is constant for all trials.*

4. *The number of trials is fixed.*

Therefore, we must be careful that four defining conditions hold good before applying the binomial distribution.

Then, the mean, $E(X^2)$, and variance for the binomial distribution is given by

$$E(X) = np, \tag{4.4.4}$$

$$E(X^2) = n(n - 1)p^2 + np,$$

and

$$Var(X) = np(1 - p). \tag{4.4.5}$$

Also,

$$\sum_{x=0}^{n} p(x) = \sum_{x=0}^{n} \binom{n}{x} p^x (1-p)^{n-x} = (q+p)^n = 1. \qquad (4.4.6)$$

Theorem 4.4.1. *Let $X_1, X_2, \ldots, X_k$ be k independent binomial variates with PDF $Bin(n_i, p)$, $i = 1, 2, \ldots, k$. Then $\sum_{i=1}^{k} X_i$ is distributed as $Bin\left(\sum_{i=1}^{k} n_i, p\right)$.*

■ Example 1

The DNA sequence of an organism consists of a long sequence of nucleotides adenine, guanine, cytosine, and thymine. It is known that in a very long sequence of 100 nucleotides, 22 nucleotides are guanine. In a sequence of 10 nucleotides, what is the probability that

(a) exactly three nucleotides will be guanine?

(b) at least four nucleotides will be guanine?

Solution

The probability p that the DNA sequence will contain guanine is

$$p = \frac{22}{100} = 0.22, \quad \text{and} \quad q = 1 - q = 0.78.$$

The probability that out of 10 nucleotides, there will be x guanine is given by

$$P(X = x) = \binom{10}{x} (0.22)^x (0.78)^{10-x}.$$

a. $P(\text{exactly three guanine}) = P(X = 3)$

$$= \binom{10}{3} (0.22)^3 (0.78)^{10-3} = 0.2244.$$

b. $P(\text{at least four guanine}) = P(X = 4) + P(X = 5) + \cdots + P(X = 10)$

$$= \sum_{x=4}^{10} P(X = x) = 1 - \sum_{x=0}^{3} P(X = x)$$

$$= 1 - [p(0) + p(1) + p(2) + p(3)]$$

$$= 0.158674.$$

The example is based on the assumption that all the nucleotides are occurring randomly in a DNA sequence. Practically, it may not be true because it is known that a nucleotide sequence shows some dependencies. This will violate the assumption of independence of each trial required for applying the binomial distribution.

Following is the R-code for example 1.

```
> #Example 1
> #Binomial Distribution
> rm(list = ls( ))
> #Part A, n = 10, p = 0.22, x = 3
> n<-10
> p<-0.22
> x<-3
> dbinom(x,n,p)
[1] 0.2244458
>
> #Part B: n = 10, p = 0.22, x = at least 4
> x1<-3
>
> p1<-1-pbinom(3,n,p)
> p1
[1] 0.1586739
>
```

∎

4.5 POISSON DISTRIBUTION

Poisson distribution is a limiting case of the binomial distribution under the following conditions:

1. The number of trials, n, is very large (that is $n \to \infty$).

2. The probability of success, p, at each trial is very small (that is $p \to 0$).

3. $np = \lambda$ is a constant.

Definition *A random variable X is said to follow a Poisson distribution if*

$$P(X = x) = \begin{cases} \dfrac{e^{-\lambda}\lambda^x}{x!}, & x = 0, 1, 2, \ldots, \\ 0, & \text{otherwise.} \end{cases} \qquad (4.5.1)$$

Here λ is a parameter of the distribution and $\lambda > 0$. We will use the notation $X \sim P(\lambda)$ to denote that the random variable X is a Poisson variate with parameter λ. Also, for any Poisson random variable X, $E(X) = \lambda$, and $\text{Var}(X) = \lambda$. Thus, mean and variance are the same in the Poisson distribution.

Poisson distribution occurs when there are infinite numbers of trials, and events occur at random points of time and space. Our interest lies only in the occurrences of the event, not in its nonoccurrences.

The events would be called to form a Poisson process if:

a. *Occurrence of any event in a specified time interval is independent of the occurrence of any event in any other time interval such that two time intervals under consideration do not overlap.*

b. *The probability that one event occurs in the sufficiently small interval $(t, t + \Delta), \Delta > 0$ is $\lambda\Delta$ where λ is a constant and Δ is small increment in time t.*

c. *The probability of occurrences of more than one event in the interval $(t, t + \Delta), \Delta > 0$ is very small.*

d. *The probability of any particular event in the time interval $(t, t + \Delta), \Delta > 0$ is independent of the actual time t, and also of all previous events.*

There are many situations in which Poisson distribution may be successfully employed: for example, number of faulty bulbs in a package, number of telephone calls received at the telephone board in some unit of time, and number of particles emitted by a radioactive material.

4.5.1 Properties of Poisson Distribution

1. **Additive Property**
   The sum of independent Poisson variates is also a Poisson variate. Thus, if $X_i, i = 1, 2, \ldots, n$ are n independent Poisson random variates with parameters $\lambda_i, i = 1, 2, \ldots, n$, respectively, then $\sum_{i=1}^{n} X_i$ is also distributed as a Poisson variate with parameter $\sum_{i=1}^{n} \lambda_i$.

2. **Conditional Distribution**
   If X and Y are independent Poisson variates with parameters $\lambda_i, i = 1, 2$, then the conditional distribution of X given $X + Y = n$ is the binomial distribution with parameters n, and $p = \frac{\lambda}{\lambda + \mu}$.

3. **Mean and Variance**
   Mean and variance both are equal to λ.

■ Example 1

An organism has two genes that may or may not express in a certain period of time. The number of genes expressing in the organism in an interval of time is distributed as Poisson variates with mean 1.5. Find the probability that in an interval

(i) Neither gene expressed.

(ii) Both genes expressed.

Solution

Since mean $\lambda = 1.5$, using (4.5.1), we find that the probability that in x genes expressed in an interval is given by

$$P[X = x] = \frac{e^{-1.5}(1.5)^x}{x!}, \quad x = 0, 1, 2, \dots$$

(i) The probability that in an interval neither genes expressed is given by

$$P(X = 0) = e^{-1.5}$$
$$= 0.2231.$$

(ii) The probability that in an interval both genes expressed is given by

$$P(X = 2) = \frac{e^{-1.5}(1.5)^2}{2!}$$
$$= 0.2510.$$

Following is the R-code for example 1:

```
> #Example 1
> #Poisson Distribution
> rm(list = ls( ))
> #Part A, lambda1 = 1.5, x = 0
>
> lambda1<-1.5
> x<-0
>
> dpois(x,lambda1)
[1] 0.2231302
>
> #Part B, x = 2
> x<-2
>
> dpois(x,lambda1)
[1] 0.2510214
>
```

4.6 NEGATIVE BINOMIAL DISTRIBUTION

Suppose for a given sequence of independent Bernoulli trials, we are interested in getting exactly r successes, where $r \geq 1$ is a fixed integer. Let p be the probability of success in a trial preceding the rth success in $x + r$ trials.

Let $NBD(x; r, p)$ denote the probability that there are x failures preceding the rth success in $x + r$ trials. Then, the last trial must be a success, which has the probability p. Thus, in $(x + r - 1)$ trials, we must have $(r - 1)$ successes whose probability is given by

$$\binom{x + r - 1}{r - 1} p^{r-1} q^x.$$

Therefore, $NBD(x; r, p)$ is given by

$$NBD(x; r, p) = \binom{x + r - 1}{r - 1} p^{r-1} q^x p$$

$$= \binom{x + r - 1}{r - 1} p^r q^x, \quad x = 0, 1, 2, \ldots; q = 1 - p. \quad (4.6.1)$$

Since

$$\binom{x + r - 1}{r - 1} = (-1)^x \binom{-r}{x},$$

therefore,

$$P(X = x) = NBD(x; r, p) = \begin{cases} \binom{-r}{x} p^r (-q)^x, & x = 0, 1, 2, \ldots, \\ 0, & \text{otherwise.} \end{cases} \quad (4.6.2)$$

Thus, $NBD(x; r, p)$ is nothing but the $(x + 1)$th term in the expansion of $p^r(1 - q)^{-r}$, which is a binomial expansion with negative index. Hence, the distribution is known as the negative binomial distribution.

The mean of the negative binomial distribution is $\frac{r}{p}$, whereas the variance is $\frac{r(1-p)}{p^2}$. Thus, we find that *mean* < *variance* in the negative binomial distribution.

■ Example 1

An organism has a large number of genes, which may or may not express under a particular condition. From previous experience, it is known that 5% of the genes may express in a particular condition. A scientist decides to examine genes randomly. What is the probability that at least five genes need to be examined to see two genes that are expressed under the particular situation?

Solution

We find that

$$r = \text{number of successes needed} = 2,$$

$$x = \text{number of failures prior to getting two successes},$$

$$p = \text{probability of success in each trial} = 0.05.$$

Thus,

$P(\text{at least five trials needed for two successes})$

$$= P(X = 5) + P(X = 6) + P(X = 7) + \ldots$$

$$= \sum_{x=5}^{\infty} \binom{x-1}{2-1} (0.05)^2 (0.95)^{x-2}$$

$$= 1 - \sum_{x=0}^{4} \binom{x-1}{2-1} (0.05)^2 (0.95)^{x-2}$$

$$= 0.98135.$$

Therefore, the probability is 0.98135 that five genes will be required to be examined before finding two genes that are expressed under the particular situation.

Following is the R-code for example 1.

```
> #Example 1
> #Negative Binomial Distribution
> rm(list = ls( ))
> # x = number of failures prior to getting r successes
> # r = number of successes needed
> # p = probability of successes in each trial
>
> x<-4
> r<-2
> p<-0.05
>
>
> p1<-1-pnbinom(x,r,p)
> p1
[1] 0.9672262
>
```

4.7 GEOMETRIC DISTRIBUTION

In a series of independent Bernoulli trials, let p be the probability of success in each trial. Suppose we are interested in getting the first success after x failures. The probability that there are x failures preceding the first success is given by

$$P(X = x) = \text{GD}(x; p) = \begin{cases} q^x p, & x = 0, 1, 2 \ldots, \\ 0, & \text{otherwise.} \end{cases} \qquad (4.7.1)$$

If we take $r = 1$ in (4.6.1), we get (4.7.1); therefore, the negative binomial distribution may be regarded as the generalization of geometric distribution.

Here

$$E(X) = \frac{1}{p}. \qquad (4.7.2)$$

$$\text{Var}(X) = \frac{1 - p}{p^2}. \qquad (4.7.3)$$

■ Example 1

A scientist is trying to produce a mutation in the genes of *Drosophila* by treating them with ionizing radiations such as $\alpha-, \beta-, \gamma-,$ or X-ray, which induces small deletions in the DNA of the chromosome. Assume that the probability of producing mutation at each attempt is 0.10.

a. What is the probability that the scientist will be able to produce a mutation at the third trial?

b. What is the average number of attempts the scientist is likely to require before getting the first mutation?

Assume that for each experiment the scientist takes a new DNA sample of *Drosophila*.

Solution

a. Let x be the number of failures before the first success. Since the first success occurs at the third trial, $x = 2, p = 0.10, q = 1 - p = 0.90$. Therefore,

$$P(\text{mutation occurs at the third trial}) = q^x p$$

$$= (0.90)^2 (0.10)$$

$$= 0.081.$$

b. The average number of attempts the scientist is likely to require before getting the first mutation is given by

$$E(X) = \frac{1}{0.10} = 10.$$

Thus, the scientist needs to have 10 trials on the average before getting the first mutation.

Following is the R-code for example 1.

```
> #Example 1
> #Geometric Distribution
> rm(list = ls( ))
> # x = number of failures prior to getting first success
> # p = probability of successes in each trial
>
> p<-0.10
> #Part A, x = 2
> x<-2
> dgeom(x,p)
[1] 0.081
>
> #Part B
> ex<-1/0.10
> ex
[1] 10
>
> p1<-1-pnbinom(x,r,p)
> p1
[1] 0.9672262
>
```

■

4.7.1 Lack of Memory

The geometric distribution has a special property known as *memoryless* property, which means that the probability that it takes an additional $X = k$ trials to obtain the first success remains unaffected by the fact that many failures have already occurred.

An event E occurs at one of the times $t = 0, 1, 2, \ldots$ and the occurrence time X has a geometric distribution.

Thus, the distribution of X is given by

$$P(X = t) = q^t p, \quad t = 0, 1, 2, \ldots \tag{4.7.4}$$

Suppose it is given that the event E has not occurred before the time k; that is, $X > k$. Let $Y = X - k$. Therefore, Y is the amount of additional time needed for the event E to occur. Then it can be shown that

$$P(Y = t | X \geq k) = P(X = t) = pq^t. \tag{4.7.5}$$

Since (4.7.4) and (4.7.5) are essentially the same equations, this means that the additional time to wait for the event E to occur has the same distribution as the initial time to wait.

4.8 HYPERGEOMETRIC DISTRIBUTION

Let us consider an urn containing N balls. Out of these N balls, M balls are white, and $N - M$ balls are red. If we draw a random sample of n balls at random without replacement from the urn, then the probability of getting x white balls out of n, $(x < n)$, is given by hypergeometric distribution. The PDF of the hypergeometric distribution is given by

$$HG(x; n, N, m) = \frac{\binom{M}{x}\binom{N-M}{n-x}}{\binom{N}{n}}; \; x = 0, 1, 2, \ldots, \min(n, m);$$

$$x \leq M \text{ and } n - x \leq N - M. \tag{4.8.1}$$

Here

$$E(X) = \frac{nM}{N} \tag{4.8.2}$$

and

$$\text{Var}(X) = \frac{nM(N - M)(N - n)}{N^2(N - 1)}. \tag{4.8.3}$$

In the binomial distribution, events are independent, but in the hypergeometric distribution, the events are dependent, since the sampling is done without replacement so that the events are stochastically dependent, although events are random.

■ Example 1

Suppose in a microarray experiment, there are 120 genes under consideration, and only 80 genes out of 120 genes are expressed. If six genes are randomly selected for an in-depth analysis, find the probability that only two of the six genes selected will be expressed genes.

Solution

Here $x = 2, n = 6, N = 120$, and $M = 80$.

Thus,

P (only two of the six genes will be expressed genes)

$$= \text{HD}(2; 6, 120, 80) = \frac{\binom{80}{2}\binom{40}{4}}{\binom{120}{6}}$$

$$= 0.079.$$

When N is large and n is relatively small as compared to N, then the hypergeometric distribution can be approximated by the binomial distribution with the parameters n, and $p = M/N$.

Following is the R-code for example 1.

```
> #Example 1
> #Hypergeometric Distribution
> rm(list = ls( ))
> # x = number of expressed genes needed from the
> # sample which is taken for an in-depth analysis
> # n = sample size taken for in-depth analysis
> # N = Total number of genes under consideration
> # M1 = Number of unexpressed genes among N genes
> # M = Number of expressed genes among N genes
>
> x<-2
> n<-6
> M1<-40
> M<-80
> dhyper(x,M,M1,n)
[1] 0.07906174
>
```

4.9 MULTINOMIAL DISTRIBUTION

Multinomial distribution can be regarded as a generalization of binomial distribution when there are n possible outcomes on each of m independent trials, with $n \geq 3$.

Let $E_1, E_2, \ldots, E_k$ be k mutually exclusive and exhaustive outcomes of a trial with the respective probabilities $p_1, p_2, \ldots, p_k$.

The probability that E_1 occurs x_1 times, E_2 occurs x_2 times, and E_k occurs x_k times in n independent trials is given by

$$\text{MND}(X_1 = x_1, X_2 = x_2, \ldots, X_k = x_k)$$

$$= \frac{n!}{x_1!\, x_2!\ldots x_k!} \cdot p_1^{x_1} p_2^{x_2} \ldots p_k^{x_k}, 0 \le x_i \le n$$

$$= \frac{n!}{\prod_{i=1}^{k}(x_i!)} \cdot \prod_{i=1}^{k} p_i^{xi}, \quad \text{where} \sum_{i=1}^{k} x_i = n. \qquad (4.9.1)$$

The term *multinomial* came from the fact that the equation (4.9.1) is the general term in the multinomial expansion

$$(p_1 + p_2 + \cdots + p_k)^n, \sum_{i=1}^{k} p_i = 1.$$

For each X_i, we have

$$E(X_i) = np_i,$$

and

$$\text{Var}(X_i) = np_i(1 - p_i).$$

■ Example 1

In an induced mutation, mutation occurs as a result of treatment with a mutagenic agent such as ionizing radiation, nonionizing radiation, or a chemical substance. It is assumed that for a randomly selected gene, the mutation in the gene will occur due to ionizing radiation, nonionizing radiation, and chemical substance with respective probabilities 2/9, 1/5, and 26/45. What is the probability that out of seven randomly selected genes, two genes will have mutation due to ionizing radiations, two genes will have mutation due to nonionizing radiations, and three genes will have mutation due to chemicals?

Solution

We find that

$$x_1 = 2, x_2 = 2, x_3 = 3, p_1 = \frac{2}{9}, p_2 = \frac{1}{5}, p_3 = \frac{26}{45}.$$

We use the multinomial distribution to find the probability that two genes will have mutation due to ionizing radiations, two genes will have mutation due to nonionizing radiations, and three genes will have mutation due to chemicals, and is given by

$$\text{MND}(x_1, x_2, x_3, p_1, p_2, p_3) = \text{MND}\left(2, 2, 3, \frac{2}{9}, \frac{1}{5}, \frac{26}{45}\right).$$

$$= \frac{7!}{2!\,2!\,3!}\left(\frac{2}{9}\right)^2\left(\frac{1}{5}\right)^2\left(\frac{26}{45}\right)^3.$$

$$= 0.08.$$

Following is the R-code for example 1.

```
> #Example 1
> #Multinomial Distribution
> rm(list = ls( ))
> # x1 = vector of genes having mutation due to ionizing radiations,
> # genes having mutation due to nonionizing radiations,
> # genes having mutation due to chemicals
>
> # p1 = vector of corresponding probabilities for x1
>
> x1<-c(2,2,3)
> p1<-c(2/9,1/5,26/45)
>
> dmultinom(x = x1, prob = p1)
[1] 0.08000862
```

■

■ Example 2

A completely recessive gene has been attributed to the absence of legs in cattle ("amputated, aa"). A normal bull (Aa) is mated with a normal cow (Aa).

P:	Aa	X	Aa
(normal)		(normal)	
F_1:	Genotypes		Phenotypes

$$\left.\begin{array}{l}\dfrac{1}{4}\ \text{AA}\\[1.5ex]\dfrac{1}{2}\ \text{Aa}\end{array}\right\} \qquad \dfrac{3}{4}\ \text{Normal}$$

$$\dfrac{1}{4}\ \text{aa} \qquad\qquad \dfrac{1}{4}\ \text{Amputated}$$

In an experiment, six offspring of normal parents were studied. What is the chance of finding three offspring of type AA, two of type Aa, and one of type aa?

Solution

Let

$$p_1 = \text{probability of offspring being type AA} = \frac{1}{4},$$

$$p_2 = \text{probability of offspring being type Aa} = \frac{1}{2},$$

$$p_3 = \text{probability of offspring being type aa} = \frac{1}{4}.$$

Let

$$x_1 = \text{number of offspring of type AA required} = 3,$$

$$x_2 = \text{number of offspring of type Aa required} = 2,$$

$$x_3 = \text{number of offspring of type aa required} = 1.$$

Then,

P (three offspring of type AA, two offspring of type Aa, and one offspring of type aa)

$$= \text{MND}(x_1, x_2, x_3, p_1, p_2, p_3) = \text{MND}\left(3, 2, 1, \frac{1}{4}, \frac{1}{2}, \frac{1}{4}\right)$$

$$= \frac{6!}{3!\,2!\,1!} \left(\frac{1}{4}\right)^3 \left(\frac{1}{2}\right)^2 \left(\frac{1}{4}\right)^1$$

$$= 0.058.$$

Following is the R-code for example 2.

```
> #Example 2
> #Multinomial Distribution
> rm(list = ls( ))
> # x1 = vector of genes having mutation due to ionizing radiations,
> # genes having mutation due to nonionizing radiations,
> # genes having mutation due to chemicals
>
> # p1 = vector of corresponding probabilities for x1
>
> x1<-c(3,2,1)
> p1<-c(1/4,1/2,1/4)
>
> dmultinom(x = x1, prob = p1)
[1] 0.05859375
>
```

■

Continuous Probability Distributions

Some of the important distributions are associated with continuous random variables and these distributions occur most prominently in statistical theory and applied frequently in bioinformatics.

4.10 RECTANGULAR (OR UNIFORM) DISTRIBUTION

A random variable X is said to have a continuous uniform distribution if its probability density function is given by

$$u(x; \alpha, \beta) = \begin{cases} \dfrac{1}{\beta - \alpha}, & \alpha < x < \beta, \\ 0, & \text{otherwise.} \end{cases}$$

The parameters α and β are constant and $\alpha < \beta$.

The mean and variance of the uniform distribution are given by

$$\mu = \frac{\alpha + \beta}{2},$$

and

$$\sigma^2 = \frac{1}{12}(\beta - \alpha)^2.$$

■ Example 1

The hydrogen bonds linking base pairs together are relatively weak. During DNA replication, two strands in the DNA separate along this line of weakness. This mode of replication, in which each replicated double helix contains one original (parental) and one newly synthesized strand (daughter), is called a semiconservative replication. Suppose in an organism, semiconservative replication occurs every half hour. A scientist decides to observe the semiconservative replication randomly. What is the probability that the scientist will have to wait at least 20 minutes to observe the semiconservative replication?

Solution

Let the random variable X denote the waiting time for the next semiconservative replication. Since semiconservative replication occurs every half hour, X is distributed uniformly on $(0, 30)$, with PDF

$$u(x; 0, 30) = \begin{cases} \dfrac{1}{30}, & 0 < x < 30, \\ 0, & \text{otherwise.} \end{cases}$$

The probability that the scientist has to wait at least 20 minutes is given by

$$P(X \geq 20) = \int_{20}^{30} f(x)dx = \frac{1}{30} \int_{20}^{30} 1 \cdot dx = \frac{1}{30}(30 - 20) = \frac{1}{3}.$$

Following is the R-code for example 1.

```
> #Example 1
> #Uniform Distribution
> rm(list = ls( ))
> # min = lower limit of replication time
> # max = upper limit of replication time
> # lower = minimum wait time for the scientist
> # upper = maximum wait time for the scientist
>
> min<-0
> max<-30
> lower<-20
> upper<-30
> p1<-punif(upper,min,max) – punif(lower,min,max)
> p1
[1] 0.3333333
```

■

Theorem 4.10.1. *If X is a random variable with a continuous distribution function F(X), then F(X) has uniform distribution on* [0, 1].

This theorem has been used to generate random numbers from different distributions and is used in empirical theory.

4.11 NORMAL DISTRIBUTION

The normal distribution is considered a cornerstone of modern statistical theory. The normal distribution was first discovered in 1733 by de Moivre (1667–1754), who obtained this continuous distribution as a limiting case of the binomial distribution. The normal distribution was later generalized by Laplace (1749–1829) who brought it to the full attention of the mathematical world by including it in his book. Due to an historical error, the normal distribution was credited to Karl Gauss (1777–1855), who first referred to it at the beginning of the 19th century as the distribution of errors of measurements involved in the calculation of orbits of heavenly bodies in astronomy. During the 18th and 19th centuries, efforts were made to establish the normal model as the underlying law ruling all continuous random variables. This led to the name of this distribution as "normal" distribution. But due to false premises, these efforts did not succeed. Pearson (1920, page 25) wrote, "Many years ago I called the Laplace-Gaussian curve the NORMAL curve, which name, while it avoids an international question of priority, has the disadvantage of leading people to believe that all other distributions of frequency are in one sense or another 'abnormal.' That belief is, of course, not justifiable." The normal model has wide applications and various properties that made it the most important probability model in statistical analysis.

Definition *A random variable X is said to have a normal distribution with parameters μ and σ^2 if its density function is given by*

$$f(x; \mu, \sigma) = \frac{1}{\sigma\sqrt{2\pi}} \exp\left[-\frac{1}{2}\left\{\frac{x-\mu}{\sigma}\right\}^2\right],$$

$$-\infty < x < \infty;$$

$$-\infty < \mu < \infty; \sigma > 0.$$

Generally, it is denoted by $X \sim N(\mu, \sigma^2)$.

Also,

$$E(X) = \mu,$$

and

$$\text{Var}(X) = \sigma^2.$$

If $X \sim N(\mu, \sigma^2)$, then $Z = \frac{x-\mu}{\sigma}$ is a standard normal variate with $E(Z) = 0$ and $\text{Var}(Z) = 1$ and we write $Z \sim N(0, 1)$.

The PDF of the standard normal variate Z is given by

$$\phi(z) = \frac{1}{\sqrt{2\pi}} e^{\frac{z^2}{2}}, -\infty < z < \infty.$$

The corresponding cumulative distribution function, denoted by $\Phi(z)$, is given by

$$\Phi(z) = \int_{-\infty}^{z} \phi(u)du = \frac{1}{\sqrt{2\pi}} \int_{-\infty}^{z} e^{\frac{u^2}{2}} du.$$

4.11.1 Some Important Properties of Normal Distribution and Normal Probability Curve

1. The curve is bell shaped and symmetrical about the line $x = \mu$.

2. Mean, median, and mode of the distribution are the same.

3. The linear combination of n independent normal variates is a normal variate. Thus, if $X_i, i = 1, 2, \ldots, n$ are n independent normal variates

with mean μ_i and variance σ_i^2 respectively, then

$$\sum_{i=1}^{n} a_i X_i \sim N\left(\sum_{i=1}^{n} a_i, \mu_i, \sum_{i=1}^{n} a_i^2, \mu_i^2\right),$$

where $a_1, a_2, \ldots, a_n$ are constant.

4. The x-axis is a horizontal asymptote to the normal curve.

5. The points of inflection of the normal curve, where the curve shifts its direction, are given by

$$[\mu \pm \sigma].$$

6. The probabilities of a point lying within 1σ, 2σ, and 3σ from the population mean are given by

$$P(\mu - \sigma < X < \mu + \sigma) = 0.6826,$$
$$P(\mu - 2\sigma < X < \mu + 2\sigma) = 0.9544,$$
$$P(\mu - 3\sigma < X < \mu + 3\sigma) = 0.9973.$$

7. Most of the known distributions such as binomial, Poisson, hypergeometric, and other distributions can be approximated by the normal distribution.

8. Many sampling distributions such as the student's t, Snedecor F, Chi-square distributions, etc., tend to normal distribution for large sample sizes.

9. If a variable is not normally distributed, it may be possible to convert it into a normal variate using transformation of a variable.

10. The entire theory of small sample tests—namely, t, F, and ψ^2 tests—is based on the assumption that the parent population from which samples are drawn is normal.

■ Example 1

Let the random variable X denote the mutations among progeny of males that received a gamma radiation of 2500 roentgen (r) units. Let X be normally distributed with a mean of 54 mutations per thousand and SD of 5 mutations per thousand.

    a. Find the probability that the number of mutations among progeny of males will be more than 65.

b. Find the probability that the number of mutations among progeny of males will be between 47 and 62.

Solution

Here $X \sim N(54, 5^2)$.

a. $P(X \geq 65) = P\left(\frac{X - \mu}{\sigma} \geq \frac{65 - 54}{5}\right) = P(Z \geq 2.2)$

$= 1 - P(Z \leq 2.2) = 1 - 0.9861 = 0.0139.$

b. $P(47 \leq X \leq 62) = P\left(\frac{47 - 54}{5} \leq \frac{X - \mu}{\sigma} \leq \frac{62 - 54}{5}\right)$

$= P(-1.4 \leq Z \leq 1.6)$

$= P(Z \leq 1.6) - P(Z \leq -1.4)$

$= 0.9542 - 0.0808$

$= 0.8644.$

Following is the R-code for example 1.

```
> #Example 1
> #Normal Distribution
> rm(list = ls( ))
> # mean = mean of the normal distribution
> # sd = standard deviation of the normal distribution
> # x = number of mutations among progeny males
>
> mean<-54
> sd<-5
>
> #Part A
> #p1 = probability that number of mutations
> # among progeny of males will be more than 65
> x<-65
> p1<-1-pnorm(x, mean, sd)
> p1
[1] 0.01390345
>
> # Part B
> #(x1, x2): number of mutations among progeny of males
>
> x1<-47
> x2<-62
>
```

```
> # p2 = probability that number of mutations
> # among progeny of males will be between (x1, x2)
>
> p2<-pnorm(x2, mean, sd)- pnorm(x1, mean, sd)
> p2
[1] 0.864444
```

■

■ Example 2

Suppose that in a microarray experiment, a gene is said to be expressed if the expression level is between 500 and 600 units. It is supposed that gene expressions are normally distributed with mean 540 units and variance 25^2 units. What is the probability that a randomly selected gene from this population will be an expressed gene?

Solution

Let X denote expression level of a gene. Thus, $X \sim N(540, 25^2)$. Probability that a gene will be called an expressed gene

$$= \text{Probability that the gene expression level is between 500 and 600 units}$$

$$= P(500 \leq X \leq 600)$$

$$= P\left(\frac{500 - 540}{25} \leq \frac{X - \mu}{\sigma} \leq \frac{600 - 540}{25}\right)$$

$$= P(-1.6 \leq Z \leq 2.4)$$

$$= P(Z \leq 2.4) - P(Z \leq -1.6)$$

$$= 0.9918 - 0.0548$$

$$= 0.937.$$

Thus, there is a 93.7% chance that the randomly selected gene will be an expressed gene.

Following is the R-code for example 2.

```
> #Example 2
> #Normal Distribution
> rm(list = ls( ))
> # mean = mean of the normal distribution
> # sd = standard deviation of the normal distribution
> # x = gene expression level
>
> mean<-540
> sd<-25
```

```
>
> #p1 = probability that gene is expressed
> #(x1, x2): gene expression level
>
> x1<-500
> x2<-600
>
> p1<-pnorm(x2, mean, sd)-pnorm(x1, mean, sd)
> p1
[1] 0.9370032
```

■

■ Example 3

Let the distribution of the mean intensities of cDNA spots, corresponding to expressed genes, be normal with mean 750 and standard deviation 50^2. What is the probability that a randomly selected expressed gene has a spot with a mean intensity

(a) less than 825?

(b) more than 650?

Solution

Let X denote mean intensity of a cDNA spot and $X \sim N(750, 50^2)$.

a. Probability that a randomly selected gene has a cDNA spot with mean intensity less than 825

$$= P(X \leq 825)$$
$$= P\left(\frac{X - \mu}{\sigma} \leq \frac{825 - 750}{50}\right)$$
$$= P(Z \leq 1.5)$$
$$= 0.9332.$$

b. Probability that a randomly selected gene has a cDNA spot with mean intensity more than 650

$$= P(X \geq 650)$$
$$= P\left(\frac{X - \mu}{\sigma} \geq \frac{650 - 750}{50}\right)$$
$$= P(Z \geq -2)$$
$$= 1 - P(Z \leq -2)$$
$$= 1 - 0.0228$$
$$= 0.9772.$$

The following R-code performs operations required for example 3.

```
> #Example 3
> #Normal Distribution
> rm(list = ls( ))
> # mean = mean of the normal distribution
> # sd = standard deviation of the normal distribution
> # x = mean intensity of a spot
>
> mean<-750
> sd<-50
>
> #Part A
>
> #p1 = probability that the mean intensity of a spot
> #is less than x
>
> x<-825
> p1<-pnorm(x, mean, sd)
> p1
[1] 0.9331928
>
> #Part B
> #p2 = probability that the mean intensity of a spot
> #is greater than x
> x<-650
> p2<-1- pnorm(x, mean, sd)
> p2
[1] 0.9772499
```

■

4.11.2 Normal Approximation to the Binomial

If X is a binomial random variable with mean $\mu = np$ and variance $\sigma^2 = npq$, then the limiting distribution of

$$Z = \frac{X - np}{\sqrt{npq}},$$

is the standard normal distribution $N(0, 1)$ as $n \to \infty$, and p is not close to 0 or 1.

The normal approximation to the probabilities will be good enough if both

$$np > 5 \text{ and } nq > 5.$$

A continuity correction factor is needed if a discrete random variable is approximated by a continuous random variable. Under the correction factor, the value of x is corrected by ± 0.5 to include an entire block of probability for that value. Correct the necessary x-values to z-values using

$$Z = \frac{X \pm 0.5 - np}{\sqrt{npq}}.$$

■ Example 4

Let the probability that a gene has a rare gene mutation be 0.45. If 100 genes are randomly selected, use the normal approximation to binomial to find the probability that fewer than 35 genes will have the rare gene mutation.

Solution

Let X denote the number of genes having the rare mutation.

Here $p = 0.45, q = 1 - p = 0.55$.

Using the normal approximation to binomial, we get

$$\mu = np = 100(0.45) = 45, \text{ variance } \sigma^2 = npq = 100$$

$$(0.45)(0.55) = 24.75.$$

The probability that fewer than 35 genes will have this rare gene mutation

$$= P(X \le 35)$$

$$= P\left(\frac{X - \mu}{\sigma} \le \frac{35.5 - 45}{\sqrt{24.75}}\right)$$

$$= P(Z \le -2.01)$$

$$= 0.0222.$$

■

4.12 GAMMA DISTRIBUTION

The density function of the gamma distribution with parameters α and β is given by

$$f(x) = \begin{cases} \frac{\beta^\alpha}{\Gamma(\alpha)} e^{-\beta x} x^{\alpha - 1}; & \beta, \alpha > 0, 0 < x < \infty, \\ 0, & \text{otherwise.} \end{cases}$$

Here $\Gamma(\alpha)$ is a value of the gamma function, defined by

$$\Gamma(\alpha) = \int_0^\infty e^{-x} x^{\alpha - 1} dx.$$

Integration by parts shows that

$$\Gamma(\alpha) = (\alpha - 1)\Gamma(\alpha - 1) \quad \text{for any } \alpha > 1,$$

and therefore, $\Gamma(\alpha) = (\alpha - 1)!$ where α is a positive integer.

$$\Gamma(1) = 1 \text{ and } \Gamma\left(\frac{1}{2}\right) = \sqrt{\pi}.$$

The mean and variance of the gamma distribution are given by

$$\mu = \frac{\alpha}{\beta}, \text{ and } \sigma^2 = \frac{\alpha}{\beta^2}.$$

4.12.1 Additive Property of Gamma Distribution

If $X_1, X_2 \ldots, X_k$ are independent gamma variates with parameters (α_i, β), $i = 1, 2, \ldots, k$ then $X_1 + X_2 + \cdots + X_k$ is also a gamma variate with parameter $\left(\sum_{i=1}^{k} \alpha_i, \beta\right)$.

4.12.2 Limiting Distribution of Gamma Distribution

If $X \sim$ gamma $(\alpha, \beta = 1)$ with $E(X) = \alpha$ and $\text{Var}(X) = \alpha$, then

$$Z = \frac{X - \mu}{\sigma} \sim N(0, 1) \text{ as } \alpha \to \infty.$$

4.12.3 Waiting Time Model

The gamma distribution is used as a probability model for waiting times. Let us consider a Poisson process with mean λ and let W be a waiting time needed to obtain exactly k changes in the process, where k is the fixed positive integer. Then the PDF of W is gamma with parameters $\alpha = k$, and $\beta = \lambda$.

■ Example 1

Translation is the process in which mRNA codon sequences are converted into an amino acid sequence. Suppose that in an organism, on average 30 mRNA are being converted per hour in accordance with a Poisson process. What is the probability that it will take more than five minutes before first two mRNA translate?

Solution

Let X denote the waiting time in minutes until the second mRNA is translated; thus, X has a gamma distribution with

$$\alpha = 2, \beta = \lambda = \frac{30}{60} = \frac{1}{2}.$$

Thus,

$$P(X > 5) = \int_5^\infty \frac{\left(\frac{1}{2}\right)^2}{\Gamma(2)} e^{-\frac{1}{2}x} x^{2-1} dx = \int_5^\infty \frac{xe^{-\frac{1}{2}x}}{4} dx$$

$$= 0.287.$$

∎

■ Example 2

A gene mutation is a permanent change in the DNA sequence that makes up a gene. Suppose that in an organism, gene mutation occurs at a mean rate of $\lambda = 1/3$ per unit time according to a Poisson process. What is the probability that it will take more than five units of time until the second gene mutation occurs?

Solution

Let X denote the waiting time in minutes until the sixth mRNA is translated; thus, X has a gamma distribution with $\alpha = 2$, and $\beta = \lambda = \frac{1}{3}$.

Thus,

$$P(X > 5) = \int_5^\infty \frac{\beta^\alpha}{\Gamma(\alpha)} e^{-\beta x} x^{\alpha-1} dx$$

$$= \int_5^\infty \frac{\left(\frac{1}{3}\right)^2}{\Gamma(2)} e^{-\frac{1}{3}x} x^{2-1} dx$$

$$= \int_5^\infty \frac{xe^{-\frac{1}{3}x}}{18} dx$$

$$= 0.2518.$$

Therefore, the probability that it will take more than four units of time until the second gene mutation occurs is 0.2518.

∎

4.13 THE EXPONENTIAL DISTRIBUTION

A continuous random variate X assuming non-negative values is said to have an exponential distribution with parameter $\beta > 0$, if its probability density function is given by

$$f(x) = \begin{cases} \beta e^{-\beta x}, & x \geq 0, \\ 0, & \text{otherwise.} \end{cases}$$

The exponential distribution is a special case of gamma distribution when $\alpha = 1$.

The mean of the exponential distribution is $\mu = \frac{1}{\beta}$, and the variance of the exponential distribution is $\sigma^2 = \frac{1}{\beta^2}$.

4.13.1 Waiting Time Model

The waiting time W until the first change occurs in a Poisson process with mean λ has an exponential distribution with $\beta = \lambda$.

■ Example 1

Cell division is a process by which a cell (parent cell) divides into two cells (daughter cells). Suppose that in an organism, the cell division occurs according to the Poisson process at a mean rate of two divisions per six minutes. What is the probability that it will take at least five minutes for the first division of the cell to take place?

Solution

Let X denote the waiting time in minutes until the first cell division takes place; then X has an exponential distribution with $\beta = \lambda = \frac{2}{6} = \frac{1}{3}$.

Thus,

$$P(X > 5) = \int_5^\infty \beta e^{-\beta x} dx = \int_5^\infty \frac{1}{3} e^{-\frac{1}{3}x} dx$$

$$= 0.1889.$$

Therefore, the probability that it will take at least five minutes for the first division of the cell to take place is 0.1889.   ■

4.14 BETA DISTRIBUTION

The probability density function of a beta distribution function is given by

$$f(x) = \begin{cases} \frac{1}{B(\mu,v)}(1-x)^{v-1}x^{\mu-1}, & (\mu, v) > 0, 0 < x < 1, \\ 0, & \text{otherwise,} \end{cases}$$

where $B(\mu, v)$ is known as beta variate and is given by

$$B(\mu, v) = \frac{\Gamma(\mu + v)}{\Gamma(\mu)\Gamma(v)}.$$

The mean and variance of this distribution are given by

$$\mu = \frac{\mu}{\mu + v}, \sigma^2 = \frac{\mu v}{(\mu + v)^2(\mu + v + 1)}.$$

4.14.1 Some Results

a. If X and Y are independent gamma variates with parameters μ and v respectively, then

   i. $X + Y$ is a gamma variate with parameter $(\mu + v)$ variate; i.e., the sum of the two independent gamma variates is also a gamma variate.

   ii. $\frac{X}{Y+X}$ is a beta variate with a parameter (μ, v) variate.

b. The uniform distribution is a special case of beta distribution for $\mu = 1$, $v = 1$.

4.15 CHI-SQUARE DISTRIBUTION

If $X_i, i = 1, 2, \ldots, n$ are n independent normal variates with mean μ_i, and variance $\sigma_i^2, i = 1, 2, \ldots, n$, then

$$\chi^2 = \sum_{i=1}^{n} \left(\frac{X_i - \mu_i}{\sigma_i}\right)^2$$

is a chi-square variable with n degrees of freedom.

Chi-square can also be obtained as a special case of gamma distribution where $\alpha = \frac{n}{2}$, and $\beta = \frac{1}{2}$.

The PDF of chi-square distribution is given by

$$f(x) = \begin{cases} \frac{1}{2^{\frac{n}{2}} \Gamma(\frac{1}{2})} e^{-\frac{1}{2}x} x^{\frac{n}{2}-1}; & 0 < x < \infty, \\ 0, & \text{otherwise.} \end{cases}$$

We noticed that the exponential distribution with $\beta = 1/2$ is the chi-square distribution with two degrees of freedom.

The mean and variance of the chi-square distribution is given by $\mu = n$, $\sigma^2 = 2n$.

4.15.1 Additive Property of Chi-square Distribution

If $X_1, X_2, \ldots, X_k$ are independent chi-square variates with parameters n_i, $i = 1, 2, \ldots, k$ then $X_1 + X_2 + \cdots + X_k$ is also a chi-square variate with $\sum_{i=1}^{k} n_i$ degrees of freedom.

4.15.2 Limiting Distribution of Chi-square Distribution

The chi-square distribution tends to the normal distribution for large degrees of freedom.

Statistical Inference and Applications

5.1 INTRODUCTION

Statistical inference is an important area of statistics that deals with the estimation (point and interval estimation) and testing of hypotheses. In a typical statistical problem, we may have the following two situations associated with a random variable X:

1. The PDF $f(x, \theta)$ or PMF $p(x, \theta)$ is unknown.

2. The form of PDF $f(x, \theta)$ or PMF $p(x, \theta)$ is known, but the parameter θ is unknown, and θ may be a vector.

In the first situation, we can deal with the problem by using methods based on the nonparametric theory. In the nonparametric theory, a particular form of the distribution is not required to draw statistical inference about the population or its parameters.

In the second situation in which the form of the distribution is known but the parameter(s) is(are) unknown, we can deal with the problem by using classical theory or Bayesian theory. This type of situation is common in statistics. For example, we may have a random variable X that has an exponential distribution with PDF $\text{Exp}(\theta)$, but the parameter θ itself is unknown. In this chapter, we will consider the second situation only.

The estimation is the procedure by which we make conclusions about the magnitude of the unknown parameters based on the information provided by a sample. The estimation can be broadly classified into two categories: namely, point estimation and interval estimation. In point estimation, we find a single estimate for the unknown parameter using the sample values, while in the case of interval estimation, we form an interval in which the value of the unknown parameter may lie. The interval obtained is known as the confidence interval. Thus, a confidence interval provides an estimated range of values (based on a sample) which is likely to include an unknown population parameter. If we draw several samples from the same population and calculate confidence intervals from these samples, then a certain percentage (confidence level) of the intervals will include the unknown population parameter. Confidence intervals are calculated so that this percentage is 95%, but we can produce a confidence interval with any percentage. The width of the confidence interval indicates the uncertainty about the unknown parameter. A very wide interval may indicate less confidence about the value of the unknown parameter. Confidence intervals are generally more informative than the simple inference drawn from the tests of hypotheses because the confidence intervals provide a range of plausible values for the unknown parameter.

In some cases, we may want to decide whether the given value of a parameter is in fact the true value of the parameter. We can decide this by testing the hypothesis. This is a statistical procedure by which a decision is made about the value of the parameter, whether the given value is the true value of the parameter among the given set of values or between two given values. The procedure involves formulating two or more hypotheses and then testing their relative strength by using the information provided by the sample.

In bioinformatics, estimation and hypothesis testing are used extensively. Let's consider one situation. The BRCA1 mutation is associated with the overexpression of several genes, such as HG2885A. If we have an expression level of HG2885A measured in several subjects, we may be interested in estimating the mean expression level of gene HG2885A and form a confidence interval for the mean expression level. We may be interested in testing the hypothesis that the mean expression level is among a particular set of expression values. If we have two sets of observations, such as healthy subjects and unhealthy subjects, we may be interested in testing the hypothesis that the mean expression level of the gene HG2885A is identical in two sets of observations. This testing procedure can be extended to multisample and multivariate situations. Because we need a sample for estimation or for hypothesis testing problems, we assume that the sample drawn from the population is a random sample. This will eliminate the bias of the investigator.

5.2 ESTIMATION

A sample statistic obtained by using a random sample is known as the *estimator* and a numerical value of the estimator is called the *estimate*.

Let $X_1, X_2, \ldots, X_n$ be a random sample of size n from a population having a PDF $f(x; \theta_1, \theta_2, \ldots, \theta_n)$, where $\theta_1, \theta_2, \ldots, \theta_n$ are the unknown parameters. We consider a particular form of the PDF given by

$$f(x; \mu, \sigma^2) = \frac{1}{\sigma\sqrt{2\pi}} \cdot e^{\left\{-\frac{(x-\mu)^2}{2\sigma^2}\right\}}, -\infty < x < \infty$$

where μ and σ^2 are unknown population parameters that denote mean and variance, respectively.

Since μ and σ^2 are the unknown population parameters, our aim is to estimate them by using the sample values. We can find an infinite number of functions of sample values, called *statistics*, which may estimate one or more of the unknown parameters. The best estimate among these infinite numbers of estimates would be that estimate which falls nearest to the true value of the unknown parameter. Thus, a statistic that has the distribution concentrated as close to the true value of the unknown parameter is regarded as the best estimate.

A good estimator satisfies the following criteria:

a. Consistency,

b. Unbiasedness,

c. Efficiency, and

d. Sufficiency.

These criteria are discussed briefly in the following sections.

5.2.1 Consistency

An estimator is said to be *consistent* if the estimator converges toward the true value of the estimator as the sample size increases.

Let $\hat{\theta}_n$ be an estimator of θ obtained using a random sample $X_1, X_2, \ldots, X_n$ of size n. Then, $\hat{\theta}_n$ will be a consistent estimator of θ if and only if

$$\lim_{n \to \infty} P(|\hat{\theta}_n - \theta| < c) = 1. \qquad (5.2.1.1)$$

Informally, when the sample size n is sufficiently large, the probability that the error made with the consistent estimator will be less than the number c.

■ Example 1

Let $X_1, X_2, \ldots, X_n$ be a random sample of size n from a Poisson population with parameter θ. Then, $\hat{\theta}_n = \overline{X}$ is a consistent estimator of the parameter θ.   ■

■ Example 2

Let $X_1, X_2, \ldots, X_n$ be a random sample of size n from a normal population with parameter (μ, σ^2), where μ is known and σ^2 is unknown. Then, the sample variance S^2 is a consistent estimator of the parameter σ^2.   ■

■ Example 3

Viruses consist of a nucleic acid (DNA or RNA) enclosed in a protein coat that is known as capsid. In some cases, the capsid may be a single protein, such as in a *tobacco mosaic* virus. In some cases, the capsid may be several different proteins, such as in the *T-even bacteriophages*. A sample of 200 observations was taken to estimate the mean number of proteins in the shell forming the capsid. The sample mean is found to be five. Assuming that the distribution of the proteins in the shell forming the capsid is normal with parameters $(\mu, 1)$, find a consistent estimate of the mean number of proteins forming the capsid.

Solution

We know that for a normal distribution, the consistent estimator of μ is $\overline{X}$; therefore, the consistent estimator of the mean number of proteins forming the capsid is

$$\hat{\mu} = 5.$$ ■

5.2.2 Unbiasedness

A estimate $\hat{\theta}_n$ is said to be an *unbiased* estimator of parameter θ if

$$E(\hat{\theta}_n) = \theta.$$ (5.2.2.1)

When $\hat{\theta}_n$ is a biased estimator of θ, then the bias is given by

$$b(\theta) = E(\hat{\theta}_n) - \theta.$$ (5.2.2.2)

If $\hat{\theta}_n$ is an unbiased estimator of θ, then it does not necessarily follow that a function of $\hat{\theta}_n, g(\hat{\theta}_n)$, is an unbiased estimator of $g(\theta)$. Also, unbiased estimators are not unique; therefore, a parameter may have more than one unbiased estimator.

■ Example 4

Let $X_1, X_2, \ldots, X_n$ be a random sample of size n from the normal population with parameter (μ, σ^2), where μ is unknown and σ^2 is known. Then the sample mean $\overline{X}$ is an unbiased estimator of μ. ■

■ Example 5

Let $X_1, X_2, \ldots, X_n$ be a random sample of size n from the exponential population with PDF denoted by

$$f(x; \theta) = \frac{1}{\theta} e^{-\frac{x}{\theta}}, 0 < x < \infty. \tag{5.2.2.3}$$

Then, the sample mean $\overline{X}$ is a consistent estimator of θ. ■

■ Example 6

In yeast, some genes are involved primarily in RNA processing. In a random sample of several experiments, the following number of genes are involved.

Experiment number	1	2	3	4	5	6	7	8	9
Number of genes	120	140	145	165	125	195	210	165	156

Suppose the genes are distributed normally with parameter $(\mu, 1)$. Find the unbiased estimate of μ, the mean number of genes involved primarily in the RNA processing.

Solution

Since the population is normally distributed with mean μ, its unbiased estimator is the sample mean $\bar{x}$. Thus, the unbiased estimate of the mean number of genes involved primarily in the RNA processing is given by

$$\bar{x} = 157.88.$$

Following is the R-code for example 6:

```
> #Example 6
> x<-c(120,140,145,165,125,195,210,165,156)
> xbar<-mean(x)
> xbar
[1] 157.8889
>
```

■

5.2.3 Efficiency

In certain cases, we may obtain two estimators of the same parameters, which may be unbiased as well as consistent. For example, in case of a normal population, the sample mean $\bar{x}$ is unbiased and a consistent estimator of the population mean μ; however, the sample median is also unbiased as well as a consistent estimator of the population mean μ. Now we have to decide which one of these two estimates is the best. To facilitate in making such a decision, we need a further criterion that will enable us to choose between the estimators. This criterion is known as the *efficiency*, which is based on the variances of the sampling distribution of estimators.

Suppose that we have two consistent estimators $\hat{\theta}_1$ and $\hat{\theta}_2$ of a parameter θ. Then, $\hat{\theta}_1$ is more efficient than $\hat{\theta}_2$ if

$$\text{Var}(\hat{\theta}_1) < \text{Var}(\hat{\theta}_2) \quad \text{for all } n. \tag{5.2.3.1}$$

If an estimator $\hat{\theta}$ is unbiased and it has the smallest variance among the class of all unbiased estimators of θ, then $\hat{\theta}$ is called the minimum variance unbiased estimator (MVUE) of θ. To check whether a given estimator is MVUE, we first find whether $\hat{\theta}$ is an unbiased estimator and then verify whether under very general conditions $\hat{\theta}$ satisfies the inequality

$$\text{Var}(\hat{\theta}) \geq \frac{1}{n \cdot E\left[\left(\frac{\partial \ln f(X)}{\partial \theta}\right)^2\right]}, \tag{5.2.3.2}$$

where $f(x)$ is the PDF of X and n is the sample size. The inequality (5.2.3.2) is also known as *Cramer-Rao inequality*.

■ Example 7

Let $X_1, X_2, \ldots, X_n$ be a random sample of size n from the normal population with parameter $(\mu, 1)$, where μ is the unknown population mean. Then, the sample mean $\bar{X}$ is the MVUE of μ.   ■

■ Example 8

Let $X_1, X_2, \ldots, X_n$ be a random sample of size n from a Poisson population, $P(X = x) = \frac{e^{-\lambda} \lambda^x}{x!}, x = 0, 1, 2, \ldots, \infty$. Then, $\hat{\lambda} = \frac{\sum_{i=1}^{n} X_i}{n}$ is the MVUE. ■

■ Example 9

In an experiment, it was examined whether a set of genes, called set A, might serve as a marker of tamoxifen sensitivity in the treatment and prevention of breast cancer. The expression levels of genes in set A were examined in the breast cancers of women who received adjuvant therapy with tamoxifen (Frasor et al., 2006). A part of the sample follows:

Affy Probe ID	Tam 4h
209988_s_at	1.0
219326_s_at	1.4
201170_s_at	1.6
220494_s_at	2.0
217967_s_at	1.0
209835_x_at	1.0
212899_at	0.7
219529_at	1.0
206100_at	1.1
210757_x_at	1.0
202668_at 210827_s_at 222262_s_at	1.2
221884_at	1.6
201911_s_at 203282_at 210658_s_at	1.3
212070_at	1.4
201631_s_at	5.0
211612_s_at	1.1
212447_at	1.3

Cells were treated with 10-8M trans-hydroxytamoxifen (Tam) for 4 hours prior to harvesting of RNA for analysis on U133A Affymetrix GeneChip microarrays. Values are based on triplicate arrays and three independent samples for each treatment, and show the mean-fold change from vehicle treated control cells. Find the MVUE of the mean μ assuming that the distribution of genes is normal.

Solution

In the case of a normal population, we know that the sample mean $\overline{X}$ is the MVUE of μ. Therefore, we find that the MVU estimate of mean expression levels of genes in set A is given by

$$\hat{\mu} = \overline{x} = 1.45.$$

Following is the R-code for example 9:

```
> #Example 9
> x<-c(1.0,1.4,1.6,2.0,1.0,1.0,0.7,1.0,1.1,1.0,1.2,1.6,1.3,1.4,5.0,1.1,1.3)
> xbar<-mean(x)
> xbar
[1] 1.452941
```
■

5.2.4 Sufficiency

Sufficiency relates to the amount of information the estimator contains about an estimated parameter. An estimator is said to be sufficient if it contains all the information in the sample regarding the parameter. More precisely, the estimator $\hat{\theta}$ is a sufficient estimator of the parameter θ if and only if the conditional distribution of the random sample $X_1, X_2, \ldots, X_n$, given $\hat{\theta}$, is independent of θ.

■ Example 10

Let $X_1, X_2, \ldots, X_n$ be a random sample of size n from the normal population with parameters (μ, σ^2). Then $\sum_{i=1}^{n} X_i$ and $\sum_{i=1}^{n} X_i^2$ are sufficient estimators for μ, and σ^2, respectively.
■

■ Example 11

Let $X_1, X_2, \ldots, X_n$ be a random sample of size n from the uniform population with parameters $(0, \theta)$. Then $\max_{1 \leq i \leq n} X_i$ is a sufficient estimator for θ.
■

■ Example 12

Let $X_1, X_2, \ldots, X_n$ be a random sample of size n from a Poisson population, $P(X = x) = \frac{e^{-\lambda}\lambda^x}{x!}, x = 0, 1, 2, \ldots, \infty$. Then, $\hat{\lambda} = \frac{\sum_{i=1}^{n} X_i}{n}$ is a sufficient estimator of λ.
■

Rao-Blackwell Theorem

Let U be an unbiased estimator of θ and T be sufficient statistics for θ based on a random sample $X_1, X_2, \ldots, X_n$. Consider the function $\phi(T)$ of the sufficient statistic defined as

$$\phi(T) = E(U|T = t)$$

which is independent of θ. Then,

$$E(\phi(T)) = \theta,$$

$$\text{Var}(\phi(T)) \leq \text{Var}(U).$$

Thus, we can obtain a minimum variance unbiased estimator by using sufficient statistics.

5.3 METHODS OF ESTIMATION

There are several good methods available for obtaining good estimators. Some of the methods are as follows:

   a. Method of Maximum Likelihood

   b. Method of Minimum Variance

   c. Method of Moment

   d. Method of Least Squares

   e. Method of Minimum Chi-square

Among these methods, the method of maximum likelihood is the most common method of estimation. It yields estimators, known as maximum likelihood estimators (MLEs), that have certain desirable properties. The MLEs are consistent, asymptotically normal, and if an MLE exists, then it is the most efficient in the class of such estimators. If a sufficient estimator exists, then it is a function of the MLE.

There are two basic problems with the maximum likelihood estimation method. The first problem is finding the global maximum. The second problem is how sensitive the estimate is to small numerical changes in the data.

Let $X_1, X_2, \ldots, X_n$ be a random sample of size n from a population with density function $f(x, \theta)$. The likelihood function is defined by

$$L = \prod_{i=1}^{n} f(x_i, \theta). \tag{5.3.1}$$

Thus, to maximize L, we need to find an estimator that maximizes L for variations in the parameter. Thus, the MLE $\hat{\theta}$ is the solution, if any, of

$$\frac{\partial L}{\partial \theta} = 0 \quad \text{and} \quad \frac{\partial^2 L}{\partial \theta^2} < 0. \tag{5.3.2}$$

■ Example 1

Let $X_1, X_2, \ldots, X_n$ be a random sample of size n from the normal population with parameter $(\mu, 1)$, where μ is the unknown population mean. Then, the sample mean $\overline{X}$ is the MLE of μ.   ■

■ Example 2

Let $X_1, X_2, \ldots, X_n$ be a random sample of size n from the Poisson population, $P(X = x) = \frac{e^{-\lambda}\lambda^x}{X!}, x = 0, 1, 2, \ldots \infty$. Then, $\hat{\lambda} = \frac{\sum_{i=1}^{n} X_i}{n}$ is the ML estimator of λ.   ■

■ Example 3

The expression level of a gene is measured in a particular experiment, and a repeated experiment gives the expression levels as 75, 85, 95, 105, and 115. It is known that the expression level of this gene follows a normal distribution. Find the maximum likelihood estimate of population mean μ.

Solution

In case of the normal population, the sample mean $\overline{X}$ is the maximum likelihood estimator of μ.

Therefore, the maximum likelihood of the mean expression level, $\overline{x} = \frac{75+85+95+105+115}{5} = 95$, is the maximum likelihood estimate.   ■

5.4 CONFIDENCE INTERVALS

In point estimation, we estimated the unknown parameter θ by a statistic $T = T(X_1, X_2, \ldots, X_n)$ where $X_1, X_2, \ldots, X_n$ is a random sample from the distribution of X, which has a density $f(x; \theta), \theta \in \Omega$. The statistic T may not be the true value of the parameter. This problem is more prominent when the distribution of T is a continuous distribution because in that case $P(T = \theta) = 0$. Moreover, the point estimate does not provide any information about the possible size of the error. It does not tell how close it might be to the unknown parameter θ. Also, it does not provide clues about the magnitude of the information on which an estimate is based. The alternative to a point estimate is to form an interval of plausible values, called a *confidence interval*. For forming the confidence interval, we need to set up the confidence level, which is a measure of degrees of reliability of the interval. The width of the interval tells about the precision of an interval estimate. The smaller the interval, the more close the estimate would be to the parameter. A confidence interval of 95% confidence level implies that 95% of all samples would give an interval that would contain the unknown parameter θ, and only 5% of the samples would give an interval that would not contain the parameter θ.

Let T_1 and T_2 be two estimators of θ such that

$$P(T_1 > \theta) = \alpha_1, \tag{5.4.1}$$

and

$$P(T_2 < \theta) = \alpha_2, \qquad\qquad (5.4.2)$$

where α_1 and α_2 are constants that are independent of θ.

Then,

$$P(T_1 < \theta < T_2) = 1 - \alpha, \qquad\qquad (5.4.3)$$

where $\alpha = \alpha_1 + \alpha_2$ is a level of confidence or level of significance.

The interval (T_1, T_2) is called the confidence interval with $(1 - \alpha)$ level of confidence. T_1 and T_2 are known as lower and upper confidence limits, respectively.

The confidence intervals are not unique. We can form many confidence intervals for the same parameter with the same level of significance. Thus, to get the best interval estimate, we should try to have the length of the confidence interval as small as possible so that the estimate can be as close as possible to the unknown parameter. For a fixed sample size, the length of the confidence interval can be decreased by decreasing the confidence coefficient $1 - \alpha$, but at the same time we lose some confidence.

Some of the common levels of significance used are $\alpha = 0.1, 0.05,$ and 0.01. Their corresponding z values are as follows:

1. $\alpha = 0.10, z_{\frac{\alpha}{2}} = 1.645$.

2. $\alpha = 0.05, z_{\frac{\alpha}{2}} = 1.96$.

3. $\alpha = 0.01, z_{\frac{\alpha}{2}} = 2.58$.

The standard normal values corresponding to α can be found using statistical software or tables.

Theorem 5.4.1. *Let* $X_1, X_2, \ldots, X_n$ *be a random sample of size n from the normal population with parameter* (μ, σ^2), *where* μ *is the unknown population mean and* σ^2 *is the known population variance. Let* $\overline{X}$ *be the value of the sample mean; then*

$$\overline{X} - z_{\frac{\alpha}{2}} \cdot \frac{\sigma}{\sqrt{n}} < \mu < \overline{X} + z_{\frac{\alpha}{2}} \cdot \frac{\sigma}{\sqrt{n}} \qquad\qquad (5.4.4)$$

is a $(1 - \alpha)100\%$ *confidence interval for the population mean* μ.

■ Example 1

The expression level of a gene is measured in a particular experiment. A random sample of 30 observations gives the following mean expression:

$$95, 90, 99, 98, 88, 86, 92, 95, 97, 99, 99, 98, 95, 87, 88, 89, 91, 92, 99, 91,$$

$$92, 87, 86, 88, 87, 87, 88, 87, 97, 98$$

It is known that the expression level of the gene follows a normal distribution. Find a 95% confidence interval for the population mean expression level.

Solution

Here $\bar{x} = 92.17$, $n = 30$, and $\sigma = 4.75$. Thus, the 95% confidence interval for the population mean expression level is given by

$$92.17 - 1.96 \cdot \frac{4.75}{\sqrt{30}} < \mu < 92.17 + 1.96 \cdot \frac{4.75}{\sqrt{30}},$$

$$90.47 < \mu < 93.86.$$

Following is the R-code for example 1:

```
> #Example 1
> #Install and load package TeachingDemos
>
> x<-c(95, 90, 99, 98, 88, 86, 92, 95, 97, 99, 99, 98,
+ 95, 87, 88, 89, 91, 92, 99, 91, 92, 87, 86, 88, 87,
+ 87, 88, 87, 97, 98)
>
> stdevx<-4.75
>
> z.test(x,mu = mean(x),stdevx,conf.level = 0.95)
        One Sample Z-test
data: x
z = 0, n = 30.000, Std. Dev. = 4.750, Std. Dev. of
the sample mean = 0.867, p-value = 1
alternative hypothesis: true mean is not equal to 92.16667
95 percent confidence interval:
90.46693 93.86640
sample estimates:
mean of x
92.16667
```

■

In many cases, the sample sizes are small, and the population standard deviation is unknown. Let S be the sample standard deviation. If the random sample is from a normal distribution, then we can use the fact that

$$T = \frac{\overline{X} - \mu}{S/\sqrt{n}}$$

has a t-distribution with $(n-1)$ degrees of freedom to construct the $(1-\alpha)$ 100% confidence interval for the population mean μ.

Theorem 5.4.2. *Let $X_1, X_2, \ldots, X_n$ be a random sample of size n (n is small) from a normal population with parameters (μ, σ^2), where μ and σ^2 are unknown. Let $\overline{X}$ be the value of the sample mean and S be the sample standard deviation; then*

$$\overline{X} - t_{\frac{\alpha}{2}, n-1} \cdot \frac{S}{\sqrt{n}} < \mu < \overline{X} + t_{\frac{\alpha}{2}, n-1} \cdot \frac{S}{\sqrt{n}} \qquad (5.4.5)$$

is a $(1-\alpha)$ 100% confidence interval for the population mean μ.

■ Example 2

A scientist is interested in estimating the population mean expression of a gene. The gene expressions are assumed to be normally distributed. A random sample of five observations gives the expression levels as 75, 85, 95, 105, and 115. Find the 95% confidence interval for the population mean expression level.

Solution

Here $\bar{x} = 95, n = 5$, and $S = 15.81$. Thus, the 95% confidence interval for the mean expression level of the population is given by

$$\overline{X} - t_{\frac{\alpha}{2}, n-1} \cdot \frac{S}{\sqrt{n}} < \mu < \overline{X} + t_{\frac{\alpha}{2}, n-1} \cdot \frac{S}{\sqrt{n}},$$

$$95 - 2.776 \cdot \frac{15.81}{\sqrt{5}} < \mu < 95 + 2.776 \cdot \frac{15.81}{\sqrt{5}},$$

$$75.36 < \mu < 114.63.$$

Following is the R-code for example 2:

```
> #Example 2
>
> x<-c(75, 85, 95, 105, 115)
>
```

```
>
> t.test(x,conf.level = 0.95)
        One Sample t-test
data: x
t = 13.435, df = 4, p-value = 0.0001776
alternative hypothesis: true mean is not equal to 0
95 percent confidence interval:
   75.36757 114.63243
sample estimates:
mean of x
      95
```

■

Theorem 5.4.3. *Let $X_1, X_2, \ldots, X_m$ and $Y_1, Y_2, \ldots, Y_n$ be two independent random samples of sizes m and n from the two normal populations X and Y with parameters (μ_x, σ_x^2) and (μ_Y, σ_Y^2), respectively. Let σ_X^2 and σ_Y^2 be the known population variances for population X and population Y, respectively. Let $\overline{X}$ and $\overline{Y}$ be the sample means of population X and population Y, respectively. Then,*

$$(\overline{X} - \overline{Y}) - z_{\frac{\alpha}{2}} \cdot \sigma_p < \mu_X - \mu_Y < (\overline{X} - \overline{Y}) + z_{\frac{\alpha}{2}} \cdot \sigma_p, \tag{5.4.6}$$

is a $(1 - \alpha)100\%$ confidence interval for the difference in population means $\mu_X - \mu_Y$ where

$$\sigma_p = \sqrt{\frac{\sigma_X^2}{m} + \frac{\sigma_Y^2}{n}}$$

is the pooled standard deviation.

■ Example 3

A group of researchers is interested in identifying genes associated with breast cancer and measuring their activity in tumor cells. In a certain laboratory using microarray analysis, researchers are studying the usefulness of two different groups, or "panels," of genes in studies on early stage tumors. DNA microarray analysis was used on primary breast tumors of 117 young patients, and supervised classification was used to identify a gene expression signature strongly predictive of a short interval to distant metastases ("poor prognosis" signature) in patients without tumor cells in local lymph nodes at diagnosis (van't Veer et al., 2002). Estrogen receptor protein (ERp) and progesterone receptor protein (PRp) are measured and assumed to be normally distributed. The population standard deviations of ERp and PRp are $\sigma_X = 37$ and $\sigma_Y = 34$, respectively. Find the 90% confidence interval for the difference in the population mean expression levels.

ERp	PRp
80	80
50	50
10	5
50	70
100	80
80	80
80	50
0	0
60	80
100	10
90	70
0	0
10	5

Solution

Here $\bar{x} = 54.61$, $m = 13$, $\bar{y} = 44.61$, $n = 13$, $\sigma_X = 37$, and $\sigma_Y = 34$.
The pooled standard deviation is given by

$$\sigma_p = \sqrt{\frac{\sigma_X^2}{m} + \frac{\sigma_Y^2}{n}} = 13.93.$$

The 90% confidence interval for difference in the mean expression level of the population is given by

$$(54.61 - 44.61) - 1.645 \cdot 13.93 < \mu_X - \mu_Y < (54.61 - 44.61) + 1.645 \cdot 13.93$$

$$- 12.91 < \mu_X - \mu_Y < 32.91.$$

Following is the R-code for example 3:

```
> #Example 3
>
> two.z.test = function(m,n,xbar, ybar, sdx, sdy,conf.level) {
+ alpha = 1-conf.level
+ zstar = qnorm(1-alpha/2)
+ sdp = sqrt((((sdx)^2)/m) + (((sdy)^2)/n))
+ zstar
+ xbar - ybar + c(-zstar*sdp,zstar*sdp)
+}
> x<-c(80, 50, 10, 50, 100, 80, 80, 0, 60, 100, 90, 0, 10)
> y<-c(80, 50, 5, 70, 80, 80, 50, 0, 80, 10, 70, 0, 5)
> xbar<-mean(x)
> ybar<-mean(y)
```

```
> sdx<-37
> sdy<-34
> m<-length(x)
> n<-length(y)
>
> two.z.test(m,n,xbar, ybar,sdx, sdy, 0.90)
[1] - 12.92378 32.92378
```

■

Theorem 5.4.4. *Let $X_1, X_2, \ldots, X_m$ and $Y_1, Y_2, \ldots, Y_n$ be two independent random samples of sizes m and n from the two normal populations X and Y with parameters (μ_x, σ_X^2) and (μ_Y, σ_Y^2), respectively, where σ_X^2 and σ_Y^2 are unknown but equal population variances. Both m and n are small. Let $\overline{X}$ and $\overline{Y}$ be the sample means of population X and population Y, respectively. Let S_X^2 and S_Y^2 be the sample variances of population X and population Y, respectively. Then,*

$$(\overline{X} - \overline{Y}) - t_{\frac{\alpha}{2}, m+n-2} \cdot S_p \sqrt{\frac{1}{m} + \frac{1}{n}} < \mu_X - \mu_Y$$

$$< (\overline{X} - \overline{Y}) + t_{\frac{\alpha}{2}, m+n-2} \cdot S_p \sqrt{\frac{1}{m} + \frac{1}{n}}, \qquad (5.4.7)$$

is a $(1 - \alpha)100\%$ confidence interval for the difference in population means $\mu_X - \mu_Y$, where

$$S_p = \sqrt{\frac{(m - 1)S_X^2 + (n - 1)S_Y^2}{m + n - 2}} \qquad (5.4.8)$$

is the pooled standard deviation.

If σ_X^2 and σ_Y^2 are the unknown but unequal population variances, then the confidence interval mentioned previously is modified as follows:

$$(\overline{X} - \overline{Y}) - t_{\frac{\alpha}{2}, r} \cdot \sqrt{\frac{S_X^2}{m} + \frac{S_Y^2}{n}} < \mu_X - \mu_Y < (\overline{X} - \overline{Y}) + t_{\frac{\alpha}{2}, r} \cdot \sqrt{\frac{S_X^2}{m} + \frac{S_Y^2}{n}}, \quad (5.4.9)$$

where

$$r = \frac{\left(\frac{S_X^2}{m} + \frac{S_Y^2}{n}\right)^2}{\frac{1}{m-1}\left(\frac{S_X^2}{m}\right)^2 + \frac{1}{n-1}\left(\frac{S_Y^2}{n}\right)^2}. \qquad (5.4.10)$$

The greatest integer in r is the degrees of freedom associated with the approximate Student-t distribution.

If $X_1, X_2, \ldots, X_n$ and $Y_1, Y_2, \ldots, Y_n$ are two dependent random samples of sizes n from the normal population, then the confidence interval for the difference in population means is given by

$$\overline{D} - t_{\frac{\alpha}{2}, n-1} \cdot \frac{S_D}{\sqrt{n}} < \mu_X - \mu_Y < \overline{D} + t_{\frac{\alpha}{2}, n-1} \cdot \frac{S_D}{\sqrt{n}}, \tag{5.4.11}$$

where

$$D_i = X_i - Y_i, \quad \overline{D} = \frac{\sum_{i=1}^{n} D_i}{n}, \tag{5.4.12}$$

and

$$S_D^2 = \frac{1}{n-1} \sum_{i=1}^{n} (D_i - \overline{D})^2. \tag{5.4.13}$$

■ Example 4

To study the genetic stability of Seoul virus strains maintained under the natural environment, the nucleotide sequences of the M genome segments of Seoul virus strains (KI strains) that were isolated from urban rats inhabiting the same enzootic focus between 1983 and 1988 were compared (Kariwa et al., 1994). The sample of size 8, denoted by X, is drawn from the population in 1983, while a sample of size 10, denoted by Y, is drawn from the population in 1988.

$$X: \quad 12, 14, 16, 18, 14, 9, 16, 13$$

$$Y: \quad 11, 11, 12, 13, 13, 11, 13, 11, 12, 13$$

Assuming that the two sets of data are independent random samples from the normal populations with equal variances, construct a 95% confidence interval for the difference between the mean number of substitutions in the nucleotide sequences of the population in 1983 and 1988.

Solution

The pooled standard deviation is given by

$$S_p = \sqrt{\frac{(m-1)S_X^2 + (n-1)S_Y^2}{m+n-2}},$$

$$S_p = \sqrt{\frac{7(7.71) + 9(0.88)}{8 + 10 - 2}},$$

$$= 1.96.$$

We find that $t_{0.025,16} = 2.120$, then the 95% confidence interval, for the difference between the mean substitutions in nucleotide sequences of the population in 1983 and 1988, is given by

$$\left[(\overline{X} - \overline{Y}) - t_{\frac{\alpha}{2}, m+n-2} \cdot S_p \sqrt{\frac{1}{m} + \frac{1}{n}}, \quad (\overline{X} - \overline{Y}) + t_{\frac{\alpha}{2}, m+n-2} \cdot S_p \sqrt{\frac{1}{m} + \frac{1}{n}}\right],$$

[0.020466, 3.79534].

Since the confidence interval does not contain zero, we conclude that $\mu_X \neq \mu_Y$. Thus, the genetic stability of the Seoul virus strains did not maintain under the natural environment.

Following is the R-code for example 4:

```
> #Example 4
>
> x<-c(12, 14, 16, 18, 14, 9, 16, 13)
> y<-c(11, 11, 12, 13, 13, 11, 13, 11, 12, 13)
> x
[1] 12 14 16 18 14 9 16 13
> y
[1] 11 11 12 13 13 11 13 11 12 13
>
> t.test(x,y, conf.level = 0.95, var.equal = TRUE)
        Two Sample t-test
data: x and y
t = 2.1419, df = 16, p-value = 0.04793
alternative hypothesis: true difference in means is not equal to 0
95 percent confidence interval:
0.02055439 3.97944561
sample estimates:
mean of x mean of y
     14    12
```

∎

Theorem 5.4.5. *Let* $X_1, X_2, \ldots, X_n$ *be a random sample of sizes n from a normal population X with parameters* (μ_x, σ_x^2)*, where* μ_x *and* σ_x^2 *are unknown. Let* $S_{\overline{X}}^2$ *be the sample variances of the population X. Then,*

$$\frac{(n-1)S_{\overline{X}}^2}{\chi_{\frac{\alpha}{2}, n-1}^2} < \sigma_{\overline{X}}^2 < \frac{(n-1)S_{\overline{X}}^2}{\chi_{1\frac{\alpha}{2}, n-1}^2}$$

is a $(1 - \alpha)100\%$ *confidence interval for* σ_x^2.

∎ Example 5

The mouse has been used as a model to study development, genetics, behavior, and disease. To study normal variation of mouse gene expression *in vivo*, a 5408-clone spotted cDNA microarray is used to quantify transcript levels in the kidney, liver, and testes from each of six normal male C57BL6 mice (Pritchard et al., 2001). The following table gives the relative intensity of variable genes in the kidney. *Relative intensity* refers to the average spot intensity of the gene relative to the mean spot intensity of all expressed genes on the array.

Gene	Accession	Relative Intensity
EST	AI428899	1.3
Collagen alpha 1	AI415173	1.4
EST	AI465155	3.8
Similar to MCP	AI666732	0.9
CiSH	AI385595	0.6
BCL-6	AI528676	1.7
PTEN	AI449154	2.5
EST	AI451961	2.1
EST	AI428930	0.7

Find the 90% confidence interval for σ_x^2.

Solution

Here $n = 9$, $s_x^2 = 1.0375$ and $\bar{x} = 1.67$, and $\alpha = 0.90$; therefore, $\chi^2_{\frac{\alpha}{2}, n-1} = \chi^2_{0.05, 8} = 15.51$ and $\chi^2_{1-\frac{\alpha}{2}, n-1} = \chi^2_{0.95, 8} = 2.73$.

Thus, the 90% confidence interval for σ_x^2 is given by

$$\frac{8 \cdot 1.0375}{15.51} < \sigma_X^2 < \frac{8 \cdot 1.0375}{2.73},$$

$$0.53 < \sigma_X^2 < 3.04.$$

Following is the R-code for example 5:

```
> #Example 5
> x<-c(1.3,1.4,3.8,0.9,0.6,1.7,2.5,2.1,0.7)
>
> int.chi = function(x,alpha = 0.10){
+ n = length(x)
+ s2 = var(x,na.rm = T)
+ lw = s2*(n-1)/qchisq(1-alpha/2,n-1)
+ up = s2*(n-1)/qchisq(alpha/2,n-1)
+ c(lw, up)
+}
>
```

```
> int.chi(x)
[1] 0.5352313 3.0373594
```
∎

5.5 SAMPLE SIZE

The sample size needed to estimate mean depends on the variance associated with the random variable under observation. If the variance is zero, then we need only one observation to estimate the mean because when there is no variance in the random variable under observation, all the observations will be same.

The $(1 - \alpha)100\%$ confidence interval for the mean μ is given by

$$\overline{X} - z_{\frac{\alpha}{2}} \cdot \frac{\sigma}{\sqrt{n}} < \mu < \overline{X} + z_{\frac{\alpha}{2}} \cdot \frac{\sigma}{\sqrt{n}}. \tag{5.5.1}$$

If we need the $(1 - \alpha)100\%$ confidence interval for the mean μ to be less than that given by $\overline{X} \pm \varepsilon$, then the sample size n needed to estimate the population mean is the solution of

$$\varepsilon = \frac{z_{\frac{\alpha}{2}} \cdot \sigma}{\sqrt{n}}, \quad \text{where} \quad \Phi\left(Z_{\frac{\alpha}{2}}\right) = 1 - \frac{\alpha}{2}. \tag{5.5.2}$$

Thus,

$$n = \frac{z_{\frac{\alpha}{2}}^2 \sigma^2}{\varepsilon^2}. \tag{5.5.3}$$

∎ Example 1

Suppose we are interested in finding the number of arrays needed to achieve the mean of the specified measure at the given maximum error of the estimate (ε). We are 99% confident that the new estimate for the population mean is within 1.25 of the true value of μ. The population variance is known to be 1.35. Find the number of arrays needed to achieve the mean at the desired ε.

Solution

We have

$$n = \frac{z_{\frac{\alpha}{2}}^2 \cdot \sigma^2}{\varepsilon^2} = \frac{(2.576)^2 \cdot 1.35}{(1.25)^2} = 5.73 \cong 6.$$

Thus, we need approximately six arrays to achieve the desired ε at the 99% confidence coefficient that the new estimate for the population mean is within 1.25 of the true value of μ and the population variance is 1.35.

Following is the R-code for example 1:

```
> # Example 1
>
> sam.size = function(varx,eps,alpha){
+ n = (qnorm(alpha/2)^2)*varx/(eps^2)
+
+ c(n)
+}
>
> sam.size(1.35,1.25,0.01)
[1] 5. 732551
>
```

■

5.6 TESTING OF HYPOTHESES

The testing of hypotheses is another important part of statistical inference. In the case of estimation, we find the expected value of the unknown population parameter, which may be a single value or an interval, while in testing of the hypotheses, we decide whether the given value of a parameter is in fact the true value of the parameter. In testing of hypotheses, we formulate two or more hypotheses and test their relative strengths by using the information provided by the sample. Thus, on the basis of a sample(s) from the population, it is decided which of the two or more complementary hypotheses is true. In this section, we will concentrate on two complementary hypotheses.

We define a hypothesis as a statement about a population parameter. A simple hypothesis completely specifies the distribution, while a composite hypothesis does not completely specify the distribution. Thus, a simple hypothesis not only specifies the functional form of the underlying population distribution, but also specifies the values of all parameters. The hypothesis is further classified as a null hypothesis or an alternative hypothesis. The null and alternative hypotheses are complementary hypotheses.

Generally, the null hypothesis, denoted by H_0, is the claim that is initially assumed to be true. The null hypothesis can be considered as an assertion of "no change, effect, or consequence." This suggests that the null hypothesis should have an assertion in which no change in situation, no difference in conditions, or no improvement is claimed. The alternative hypothesis is the assertion, denoted by H_a. For example, in the U.S. judicial system, a person is assumed to be innocent until proven guilty. This means that the null hypothesis is that the person is innocent, while the alternative hypothesis is that the person is guilty. The burden of proof is always on the prosecution to reject the null hypothesis.

The form of the null hypothesis is given by

$$H_0 : \text{population parameter or form } = \text{hypothesized value,}$$

where the hypothesized value is specified by the investigator or the problem itself.

The alternative hypothesis can take one of the following three forms:

- H_a: population parameter $>$ hypothesized value.

  This form is called one-sided alternative to right.

- H_a: population parameter $<$ hypothesized value.

  This form is called one-sided alternative to left.

- H_a: population parameter $\neq$ hypothesized value.

  This form is called a two-sided alternative.

The form of the alternative hypothesis should be decided in accordance with the objective of the investigation.

A test of hypotheses or test procedure is a method to decide whether the null hypothesis is rejected or accepted on the basis of a sample datum. The testing procedure has two components: (1) a test statistic or function of the sample datum is used to make the decision; (2) the test procedure partitions the possible values of the test statistic into two subsets: an acceptance region for the null hypothesis and a rejection region for the null hypothesis. A rejection region consists of those values of the test statistic that lead to the rejection of the null hypothesis. The null hypothesis is rejected if and only if the numerical value of the test statistic, calculated on the basis of the sample, falls in the rejection region. In addition to the numerical value of the test statistic, sometimes we calculate the p-value, which is a probability calculated using the test statistic. If the p-value is less than or equal to the predecided level of significance, then we reject the null hypothesis.

The p-value is defined as the probability of getting a value for the test statistic as extreme or more extreme than the observed value given that H_0 is true.

The level of significance is defined as the probability that the test statistic lies in the critical region when H_0 is true.

In the process of testing a statistical hypothesis, we encounter four situations that determine whether the decision taken is correct or in error. These four situations are shown in Table 5.6.1.

Out of these four situations, two situations lead to wrong decisions. A type I error occurs when we reject the H_0 when in fact it is true. A type II error occurs when we accept H_0 when in fact it is false.

We would like to have a test procedure that is free from type I or type II errors, but this is possible only when we consider the entire population to make a

Table 5.6.1 Decision Table and Types of Errors

Decision Taken	True Situation	
	H_0 is true	H_0 is false
Accept H_o	Correct Decision	Type II error
Reject H_o	Type I error	Correct Decision

decision to accept or reject the null hypothesis. The procedure based on a sample may result in an unrepresentative sample due to sample variability and therefore may lead to erroneous rejection or acceptance of the null hypothesis. Since, in statistical inference, the testing procedures are based on samples drawn from the population, the errors are bound to happen. A good testing procedure is one that will have a small probability of making either type of error.

The probability of committing a type I error is denoted by α, and the probability of committing a type II error is denoted by β. The type I error α is also known as a level of significance, and $1 - \beta$ is known as the power of a test. Thus, minimizing β will maximize the power of the test. The power of the test is the probability that the test will reject the null hypothesis when the null hypothesis is in fact false and the alternative hypothesis is true. The choice of rejection region fixes both α and β. Thus, our aim is to select a rejection region from a set of all possible rejection regions that will minimize the α and β. But the way α and β are related to each other, it is not possible to minimize both of them at the same time. We can see this easily when we consider extreme cases such as when we select the rejection region $C = \phi$. In this case the null hypothesis will never be rejected, and therefore $\alpha = 0$ but $\beta = 1$. Since, in general, α is considered to be more serious in nature than β, we fix α in advance and try to minimize β. Ideally, we would like α to be small and power $(1 - \beta)$ to be large. A graph of $(1 - \beta)$ as a function of the true values of the parameter of interest is called the *power curve* of the test.

The seriousness of α can be seen in the case of trial of a person accused of murder. The null hypothesis is that the person is innocent, whereas the alternative hypothesis is that the person is guilty. If the court commits a type I error, then the innocent person will be executed (assuming the death penalty as a punishment in such cases). If the court commits a type II error, then the murderer will be set free. If a type II error is committed, the court has another chance to bring the murderer back to trial if new evidence is discovered after some time. If a type I error is committed, the court has no second chance to bring the person back to court (assuming that the accused person has already been executed) if new evidence is discovered after some time proving that the accused was innocent. Thus, in this extreme scenario, the type I error is more serious in nature than the type II error.

■ Example 1

The expression level of a gene is measured in a particular experiment. A random sample of 25 observations gives the mean expression level as 75 and the sample standard deviation as 15. It is known that the expression level follows a normal distribution. Find the power of the test when the population mean μ is actually 78, the level of significance α is 0.05, and the population mean under the null hypothesis is μ_0 is 75.5.

Solution

Here $\mu_0 = 75.5$, $n = 25$, $s = 15$, and $H_a : \mu_a = 78$.

Thus, the acceptance region is given by $\mu_0 \pm z_{\frac{\alpha}{2}} \cdot \frac{s}{\sqrt{n}}$, which is

$$\left\{ 75.5 - 1.96 \cdot \frac{15}{\sqrt{25}}, 75.5 + 1.96 \cdot \frac{15}{\sqrt{25}} \right\},$$

$$\{69.62, 81.38\}.$$

Thus,

$$\beta = P(\text{accept } H_0 \text{ when } \mu = 78)$$
$$= P\{69.62 < \bar{x} < 81.38\}$$
$$= P\left\{ \frac{69.62 - 78}{\frac{15}{\sqrt{25}}} < z < \frac{81.38 - 78}{\frac{15}{\sqrt{25}}} \right\}$$
$$= P\{-2.79 < z < 1.126\}$$
$$= 0.866.$$

Therefore, the power of the test is

$$\text{Power} = 1 - \beta$$
$$= 1 - 0.866 = 0.134.$$

In this case, the probability of committing a type II error is very high; therefore, the power of the test is very low.

Similar calculations can be done if the test is assumed to be a one-sample t-test or a two-sample test.

Following is the R-code for example 1:

```
> #Example 1
>
> #Based on normal distribution
```

```
> #n = sample size
> #conf.level = confidence level
> #mu = mean under null hypothesis
> #sdx = standard deviation
> #delta = difference between true mean and mu
>
> power.z.test = function(n,mu,sdx,delta,conf.level){
+ alpha = 1 - conf.level
+ zstar = qnorm(1 - alpha/2)
+ error = zstar*sdx/sqrt(n)
+ left<-mu-error
+ right<-mu + error
+ assumed<- mu + delta
+ Zleft<-(left-assumed)/(sdx/sqrt(n))
+ Zright<-(right-assumed)/(sdx/sqrt(n))
+ p<-pnorm(Zright)-pnorm(Zleft)
+ power = 1-p
+ c(power)
+}
> power.z.test(25,75.5,15,2.5,0.95)
[1] 0.1325580
> >
#Based on t-test
> #Based on one-sample t-test
> power.t.test(n = 25,delta = 2.5,sd = 15,sig.level = 0.05,
+ type = "one.sample",alternative = "two.sided",strict = TRUE)
One-sample t test power calculation
n = 25
delta = 2.5
sd = 15
sig. level = 0.05
power = 0.1259956
alternative = two.sided
```

The following steps are required for testing the hypothesis:

1. Identify the parameter of interest. It may be population mean, population variance, population distribution, etc. The testing of the hypothesis is always about population parameters or its distribution. It is never about a sample or sample characteristics.

2. State the null hypothesis.

3. State the alternative hypothesis.

4. State the test statistic to compute the test statistic value under the null hypothesis.

5. Choose the significance level.

6. State the rejection area and the rejection criteria. The rejection criteria may be critical value approach or p-value approach.

7. State the conclusion whether the null hypothesis is rejected or accepted.

5.6.1 Tests about a Population Mean

Different tests about a population mean can be applied based on the assumptions of the population variance, sample size, and population distribution.

Case I (Single sample case): Population is normal, σ known, sample size is large.

In this case, the population standard deviation is known. Thus, we can use a Z-test, which is also known as a standardized test. The distribution of the z-statistic is normal.

1. **Assumptions**

   The population is normal, population variance is known, and sample size is large.

2. **Hypotheses**
   Null hypothesis

   $$H_0 : \mu = \mu_0.$$

   Alternative hypotheses

   $$H_A : \mu < \mu_0,$$
   $$H_A : \mu > \mu_0,$$
   $$H_A : \mu \neq \mu_0.$$

3. **Test statistic**

   $$Z = \frac{\overline{X} - \mu_0}{\sigma / \sqrt{n}}. \qquad (5.6.1.1)$$

4. **Rejection criteria**
   Choose one or both rejection criteria.

   A. Critical value approach
   1. For $H_A : \mu < \mu_0$, reject the null hypothesis if $Z < -Z_\alpha$.

   2. For $H_A : \mu > \mu_0$, reject the null hypothesis if $Z > Z_\alpha$.

   3. For $H_A : \mu \neq \mu_0$, reject the null hypothesis if $Z < -Z_{\alpha/2}$ or $Z > Z_{\alpha/2}$.

B. *p*-value approach

Compute the *p*-value for the given alternative as follows:

1. For $H_A : \mu < \mu_0, p = P(Z < -z)$.

2. For $H_A : \mu > \mu_0, p = P(Z > z)$.

3. For $H_A : \mu \neq \mu_0, p = 2P(Z > |z|)$.

Reject the null hypothesis if $p < \alpha$.

5. **Conclusion**

Reject or accept the null hypothesis on the basis of the rejection criteria.

■ Example 2

Genes are organized in operons in the prokaryotic genome. The genes within an operon tend to have similar levels of expression. An experiment is conducted to find the true mean expression of an operon. The mean expression level of an operon consisting of 34 genes is found to be 0.20. Test the claim that the true mean expression of the population is 0.28 when the population variance is 0.14. Use the significance level as $\alpha = 0.05$.

Solution

1. **Assumptions**

   The population is normal, the population variance is known, and the sample size is large.

2. **Hypotheses**

   Null hypothesis

   $$H_0 : \mu = \mu_0 = 0.28.$$

   Alternative hypothesis

   $$H_A : \mu \neq 0.28.$$

3. **Test statistic**

   $$Z = \frac{\bar{x} - \mu_0}{\sigma / \sqrt{n}} = \frac{0.20 - 0.28}{\sqrt{0.14} / \sqrt{34}} = 1.24.$$

4. **Rejection criteria**

   A. Critical value approach

   For $H_A: \mu \neq \mu_0$, since we have $z = 1.24 < Z_{0.05/2} = 1.96$, we fail to reject the null hypothesis.

   B. *p*-value approach

   Compute the *p*-value for the given alternative as follows:

   For $H_A: \mu \neq \mu_0, p = 2P(Z > |z|) = 2 \cdot 0.1075 = 0.215$.

   Since $0.215 = p > \alpha$, we fail to reject the null hypothesis.

5. Conclusion

We fail to reject the null hypothesis. Thus, the true mean expression of the population is 0.28 when the population standard deviation is 0.14.

The following R-code illustrates these calculations:

```
> #Example 2
> #Based on normal distribution
>
> #n = sample size
> #mu = mean under null hypothesis
> #xbar = sample mean
> #varx = population variance
> #conf.level = confidence level
>
> z.test.onesample = function(n,mu, xbar, varx, alternative = "two sided", conf.level) {
alpha = 1-conf.level
p<-2*pnorm(xbar, mu, sqrt(varx/n))
print ("Two-sided alternative")
cat("p-value = ",p,fill = T)
cat("Alpha = ",alpha, fill = T)
if(p < alpha)"Null hypothesis is rejected"
else ("Decision: Since p > alpha, failed to reject Null hypothesis")
}
>
> z.test.onesample(34,0.28,0.20,0.14, alternative = "two sided", 0.95)
[1] "Two-sided alternative"
p-value = 0.2125039
Alpha = 0.05
[1] "Decision: Since p > alpha, failed to reject Null hypothesis"
>
```

■

Case II (Single sample case): Population is normal, σ unknown, and the sample is small

The Z-test statistic does not have normal distribution if the population standard deviation is unknown and the sample size is small. The standard normal Z-test for testing the single population mean can be applied even if the population variance is unknown, provided the sample size is large. In that case the sample standard deviation is used to estimate the population standard deviation. But the central limit theorem cannot be applied if the sample size is small. If the unknown population variance is estimated by the sample standard deviation and the sample size is small, the resulting test is called the Students' t-test for the population mean. (This test was called t-test by W. G. Gosset under the pen name "student.") A sample of size less than 28 is considered to be small. The test procedure is as follows:

1. **Assumptions**
   The population is normal, the population variance is unknown, and the sample size is small.

2. **Hypotheses**

   Null hypothesis
   $$H_0 : \mu = \mu_0.$$

   Alternative hypotheses
   $$H_A : \mu < \mu_0,$$
   $$H_A : \mu > \mu_0,$$
   $$H_A : \mu \neq \mu_0.$$

3. **Test statistic**

   $$t = \frac{\overline{X} - \mu_0}{S / \sqrt{n}}, \tag{5.6.1.2}$$

   with $v = n - 1$ degrees of freedom.

4. **Rejection criteria**
   Choose one or both rejection criteria.

   A. Critical value approach
   1. For $H_A : \mu < \mu_0$, reject the null hypothesis if $t < -t_{\alpha,v}$.

   2. For $H_A : \mu > \mu_0$, reject the null hypothesis if $t > t_{\alpha,v}$.

   3. For $H_A : \mu \neq \mu_0$, reject the null hypothesis if $t < -t_{\alpha/2,v}$ or $t > t_{\alpha/2,v}$.

   B. *p*-value approach
   Compute the *p*-value for the given alternative as follows:
   1. For $H_A : \mu < \mu_0, p = P(T < -t)$.

   2. For $H_A : \mu > \mu_0, p = P(T > t)$.

   3. For $H_A : \mu \neq \mu_0, p = 2P(T > |t|)$.

   Reject the null hypothesis if $p < \alpha$.

5. **Conclusion**
   Reject or accept the null hypothesis on the basis of rejection criteria.

■ Example 3

In DNA, the proportion of G is always the same as the proportion of C. Thus, the composition of any DNA can be described by the proportion of its bases, which is $G + C$. Suppose, in an experiment on particular species, the sample mean number of $G + C$ is found to be 300, while the sample standard deviation is found to be 15 in 16 DNA samples. Test the claim that this particular species has a mean number of $G + C$ bases as 310 at $\alpha = 0.05$.

Solution

1. **Assumptions**
   The population is normal, the population variance is unknown, and the sample size is small.

2. **Hypotheses**
   Let μ denote the mean number of $G +$ bases in the species.

   Null hypothesis
   $$H_0 : \mu = 300.$$

   Alternative hypothesis
   $$H_A : \mu \neq 300.$$

3. **Test statistic**
   $$t = \frac{310 - 300}{15 / \sqrt{16}} = 2.66 \text{ with } v = 16 - 1 = 15 \text{ degrees of freedom.}$$

4. **Rejection criteria**

   A. Critical value approach
   For $H_A : \mu \neq 300$, $t_{\alpha/2,v} = t_{0.025,15} = 2.131$. Since $t = 2.66 > t_{0.025,15}$, we reject the claim that the population mean number of $G + C$ bases is 300.

   B. p-value approach
   The p-value for the $H_A : \mu \neq 300$ is given by
   $$2 \cdot 0.005 < p < 2 \cdot 0.01,$$
   $$0.01 < p < 0.02.$$

   Since $p < 0.05$, we reject the claim that the population mean of $G + C$ is 300.

5. **Conclusion**
   We reject the claim that the population mean number of $G + C$ is 300 in that particular species.

The following R-code provides calculations for example 3.

```
> #Example 3
>
> #n = sample size
> #mu = mean under null hypothesis
> #xbar = sample mean
> #sdx = sample standard deviation
> #conf.level = confidence level
>
> t.test.onesample = function(n,mu, xbar, sdx, alternative = "two sided",
conf.level){
alpha = 1-conf.level
t<-(xbar-mu)*sqrt(n)/sdx
cat("t-test statistic value = ",t,fill = T)
df<-n-1
p<-2*(1-pt(abs(t), df))
print("Two-sided alternative")
cat("p-value = ",p,fill = T)
cat("Alpha =",alpha,fill = T)
if(p<alpha)"Null hypothesis is rejected"
else("Decision: Since p > alpha,failed to reject Null hypothesis")
}
>
> t.test.onesample(16,300, 310, 15, alternative = "two sided", 0.95)
t-test statistic value = 2.666667
[1] "Two-sided alternative"
p-value = 0.01759515
Alpha = 0.05
[1] "Null hypothesis is rejected"
>
```

Case III (Two-sample case): Populations are normal, population variances are known, sample sizes are large

The standard normal test for testing the difference in means can be used in this case. The distribution of Z-statistic is normal.

1. **Assumptions**

   The two populations are normal, the population variances are known, and sample sizes are large.

2. **Hypotheses**

   Let μ_1, μ_2, σ_1^2, and σ_2^2 be the mean of population 1, mean of population 2, variance of population 1, and variance of population 2, respectively.

   Null hypothesis
   $$\mu_1 - \mu_2 = \Delta,$$

   where Δ is a constant.

Alternative hypotheses

$$H_A : \mu_1 - \mu_2 < \Delta,$$
$$H_A : \mu_1 - \mu_2 > \Delta,$$
$$H_A : \mu_1 - \mu_2 \neq \Delta.$$

3. **Test statistic**

   Let $\bar{x}_1, \bar{x}_2, n_1$, and n_2 be the sample mean from population 1, sample mean from population 2, sample size from population 1, and sample size from population 2. Then the test statistic is given by

   $$Z = \frac{\bar{X}_1 - \bar{X}_2 - \Delta}{\sqrt{\sigma_1^2 / n_1 + \sigma_2^2 / n_2}}. \tag{5.6.1.3}$$

4. **Rejection criteria**

   Choose one or both rejection criteria.

   A. Critical value approach

   1. For $H_A : \mu_1 - \mu_2 < \Delta$, reject the null hypothesis if $Z < -Z_\alpha$.

   2. For $H_A : \mu_1 - \mu_2 > \Delta$, reject the null hypothesis if $Z > Z_\alpha$.

   3. For $H_A : \mu_1 - \mu_2 \neq \Delta$, reject the null hypothesis if $Z < -Z_{\alpha/2}$ or $Z > Z_{\alpha/2}$.

   B. p-value approach

   Compute the p-value for the given alternative as follows:

   1. For $H_A : \mu_1 - \mu_2 < \Delta, p = P(Z < -z)$.

   2. For $H_A : \mu_1 - \mu_2 > \Delta, p = P(Z > z)$.

   3. For $H_A : \mu_1 - \mu_2 \neq \Delta, p = 2P(Z > |z|)$.
   Reject the null hypothesis if $p < \alpha$.

5. **Conclusion**

   Reject or accept the null hypothesis on the basis of rejection criteria.

■ Example 4

The denaturation of DNA occurs over a narrow temperature range and results in striking changes in many of its physical properties. The midpoint of the temperature range over which the strands of DNA separate is called the melting temperature. The melting temperature may vary from species to species depending on the number of G and C base pairs contained in DNA. To compare the melting temperature of two species A and B, two samples of sizes 34 and 45 were taken from species A and B, respectively. It was found that the mean temperatures of species are 82°F and 85°F for species A and B, respectively. The population standard deviations of species A and B were 12°F and 14°F, respectively. Test the claim that two species have the same mean melting temperature at $\alpha = 0.05$.

Solution

1. **Assumptions**

   In this case, we assume that the two populations are normal. The population variances are known, and sample sizes are large.

2. **Hypotheses**

   We have $\sigma_1^2 = 12^2$, and $\sigma_2^2 = 14^2$.
   We set up the null hypothesis as follows:

   $$H_0 : \mu_1 - \mu_2 = 0.$$

   The alternative hypothesis is set as follows:

   Alternative hypothesis

   $$H_A : \mu_1 - \mu_2 \neq 0.$$

3. **Test statistic**

   Here $\bar{x}_1 = 82, \bar{x}_2 = 85, n_1 = 34$ and $n_2 = 45$.
   The test statistic is calculated as follows:

   $$z = \frac{\bar{x}_1 - \bar{x}_2}{\sqrt{\sigma_1^2/n_1 + \sigma_2^2/n_2}} = \frac{82 - 85}{\sqrt{12^2/34 + 14^2/45}} = -1.02.$$

4. **Rejection criteria**

   We apply both criteria.

   A. Critical value approach

   For $H_A : \mu_1 - \mu_2 \neq 0$, the $z_{\alpha/2} = z_{0.025} = 1.96$.
   Since $z = -1.02 > -z_{0.025} = -1.96$, we fail to reject the null hypothesis.

   B. p-value approach

   For $H_A : \mu_1 - \mu_2 \neq 0$,

   $$p = 2P(Z > |z|) = 2(1 - 0.8461) = 0.3078.$$

   Since $p > \alpha$, we fail to reject the null hypothesis.

5. Conclusion

We fail to reject the claim that two species have the same mean melting temperature at $\alpha = 0.05$.

The R-code for carrying out calculations for example 4 is as follows:

```
> #Example 4
> #Based on normal distribution
> #n1 = first sample size
> #n2 = second sample size
> #mu1 = mean of population 1 under null hypothesis
> #mu2 = mean of population 2 under null hypothesis
> #xbar = sample mean of population 1
> #ybar = sample mean of population 2
> #sdx = standard deviation of population 1
> #sdy = standard deviation of population 2
> #conf.level = confidence level
>
> z.test.twosample = function(n1, n2, mu1, mu2, xbar, ybar, sdx, sdy, alternative = "two sided",
conf.level){
alpha = 1-conf.level
z <-(xbar − ybar − (mu1 − mu2))/sqrt((sdx^2/n1) + (sdy^2/n2))
p <-2*(1- pnorm(abs(z)))
print("Two-sided alternative")
cr <-qnorm(1- alpha/2)
cat("p-value = ",p,"Z-test Statistic value = ",z, "Critical Value = ", cr, fill = T)
cat("Alpha = ",alpha,fill = T)
if(p<alpha)"Null hypothesis is rejected"
else("Decision: Since p>alpha,failed to reject Null hypothesis")
}
>
> z.test.twosample(34, 45, 0, 0, 82, 85, 12, 14, alternative = "two sided", 0.95)
[1] "Two-sided alternative"
p-value = 0.3060544 Z-test Statistic value = -1.023536
Critical Value = 1.959964
Alpha = 0.05
[1] "Decision: Since p > alpha, failed to reject Null hypothesis"
>
>
```

Case IV (Two-sample case): Population distribution is normal, $\sigma_1 = \sigma_2$ but unknown, sample sizes are small, and samples are independent

Since the sample sizes are small, the standard normal test cannot be applied for this case. We can apply a t-test for the independent samples. The unknown variances are estimated by corresponding sample variances. The test statistic is the same as that of the Z-test for two samples, except that the distribution of the test statistic under the null hypothesis is a t-distribution with $n_1 + n_2 - 2$ degrees of freedom, and population variances are replaced by sample variances.

1. **Assumptions**
   The two populations are normal, population variances are unknown but equal, and sample sizes are small.

2. **Hypotheses**
   Let μ_1, μ_2, σ_1^2, and σ_2^2 be the mean of population 1, mean of population 2, variance of population 1, and variance of population 2, respectively.

   Null hypothesis
   $H_0 : \mu_1 - \mu_2 = \Delta$, where Δ is a constant.

   Alternative hypotheses

   $$H_A : \mu_1 - \mu_2 < \Delta,$$
   $$H_A : \mu_1 - \mu_2 > \Delta,$$
   $$H_A : \mu_1 - \mu_2 \neq \Delta.$$

3. **Test statistic**
   Let $\overline{X}_1, \overline{X}_2, n_1$ and n_2, be the sample mean from population 1, sample mean from population 2, sample size from population 1, and sample size from population 2. Let S_1^2, and S_2^2 be the sample variances of samples drawn from population 1 and population 2, respectively. Then the test statistic is given by

   $$t = \frac{\overline{X}_1 - \overline{X}_2 - \Delta}{S_p \sqrt{\frac{1}{n_1} + \frac{1}{n_2}}}, \qquad (5.6.1.4)$$

   where $S_p = \sqrt{\frac{(n_1-1)S_1^2 + (n_2-1)S_2^2}{n_1+n_2-2}}$,
   and $v = n_1 + n_2 - 2$.

4. **Rejection criteria**
   Choose one or both rejection criteria.

   A. Critical value approach
   1. For $H_A : \mu_1 - \mu_2 < \Delta$, reject the null hypothesis if $t < -t_{\alpha,v}$.

   2. For $H_A : \mu_1 - \mu_2 > \Delta$, reject the null hypothesis if $t > t_{\alpha,v}$.

   3. For $H_A : \mu_1 - \mu_2 \neq \Delta$, reject the null hypothesis if $t < -t_{\alpha/2,v}$ or $t > t_{\alpha/2,v}$.

B. *p*-value approach

Compute the *p*-value for the given alternative as follows:

   1. For $H_A : \mu_1 - \mu_2 < \Delta, p = P(T < -t)$.

   2. For $H_A : \mu_1 - \mu_2 > \Delta, p = P(T > t)$.

   3. For $H_A : \mu_1 - \mu_2 \neq \Delta, p = 2P(T > |t|)$.

   Reject the null hypothesis if $p < \alpha$.

5. Conclusion

Reject or accept the null hypothesis on the basis of rejection criteria.

■ Example 5

One of the important functions in a cell is protein synthesis. To produce sufficient proteins, a cell uses many ribosomes contained in the cell. The number of proteins varies in different types of cells. Bacteria contain approximately 10,000 ribosomes. To test whether two species of bacteria have the same mean number of ribosomes, a sample size of 12 was drawn from each species. It was found that the sample mean and sample standard deviation of species 1 are 10,500 and 25, respectively, while for species 2, the sample mean and standard deviation are 10,512, and 50, respectively. Test whether the mean number of ribosomes is the same in two species. Use a 5% level of significance. Assume that the populations under consideration are normally distributed and population variances are unknown but equal.

Solution

1. Assumptions

The two populations are normal, population variances are unknown but equal, and sample sizes are small.

2. Hypotheses

Let μ_1, μ_2, σ_1^2, and σ_2^2 be the mean of population 1, mean of population 2, variance of population 1, and variance of population 2, respectively.

In this problem, we have

Null hypothesis

$H_0 : \mu_1 - \mu_2 = 0$, where Δ is a constant.

Alternative hypothesis

$H_A : \mu_1 - \mu_2 \neq \Delta$.

3. Test statistic

Let $\bar{x}_1, \bar{x}_2, n_1$, and n_2 be the sample mean from population 1, sample mean from population 2, sample size from population 1, and sample size from population 2. Let s_1^2 and s_2^2 be the sample variances of samples drawn from population 1 and population 2, respectively. Then, the test statistic is given by

$$t = \frac{\bar{x}_1 - \bar{x}_2 - \Delta}{s_p \sqrt{\left(\frac{1}{n_1} + \frac{1}{n_2}\right)}}, \text{ where } s_p = \sqrt{\frac{(n_1 - 1)s_1^2 + (n_2 - 1)s_2^2}{n_1 + n_2 - 2}},$$

and $v = n_1 + n_2 - 2$.
In this problem, we find that

$$\bar{x}_1 = 10{,}500, \bar{x}_2 = 10{,}512, n_1 = 12, s_1^2 = 25^2, n_2 = 12, s_2^2 = 50^2,$$

$$s_p = \sqrt{\frac{(12 - 1)625 + (12 - 1)2500}{22}} = 39.52,$$

and $v = 22$.
Thus,

$$t = \frac{10500 - 10512}{39.52\sqrt{\frac{1}{12} + \frac{1}{12}}} = -0.74377.$$

4. **Rejection criteria**

   Choose one or both rejection criteria.

   A. Critical value approach
   Here $t_{\alpha/2,v} = t_{0.025,22} = 2.074$.
   For $H_A : \mu_1 - \mu_2 \neq \Delta$,
   we find that $-0.74377 > -2.074$; therefore, we fail to reject the null hypothesis.

   B. p-value approach
   We compute the p-value as follows:
   For $H_A : \mu_1 - \mu_2 \neq \Delta, p = 2P(T > |t|) = P(T > 0.74377)$.
   Therefore, $p = 0.4726$.
   Since $p > \alpha = 0.05$, we fail to reject the null hypothesis.

5. **Conclusion**

   On the basis of the given data, we fail to reject the null hypothesis. Thus, the mean number of ribosomes is the same in two species.

The R-code for example 5 is as follows:

```
> #Example 5
>
> #n1 = first sample size
> #n2 = second sample size
> #mu1 = mean of population 1 under null hypothesis
```

```
> #mu2 = mean of population 2 under null hypothesis
> #xbar = sample mean of population 1
> #ybar = sample mean of population 2
> #sdx = standard deviation of population 1
> #sdy = standard deviation of population 2
> #conf.level = confidence level
>
> t.test.twosample = function(n1, n2, mu1, mu2, xbar, ybar, sdx, sdy, alternative =
"two sided",
conf.level) {
alpha = 1-conf.level
s<-sqrt(((nl-l)*sdx^2 + (n2-1)*sdy^2)/(n1 + n2-2))
t<-(xbar-ybar-(mu1-mu2))/(s*sqrt(1/n1 + 1/n2))
cat("t-test statistic value = ", t, fill = T)
df<-nl + n2-2
p<-2*(1-pt(abs(t), df))
print("Two-sided alternative")
cat("p-value = ",p,fill = T)
cat("Alpha = ",alpha,fill = T)
if(p < alpha)"Null hypothesis is rejected"
else ("Decision: Since p > alpha, failed to reject Null hypothesis")
}
>
> t.test.twosample(12, 12, 0, 0, 10500, 10512, 25, 50, alternative = "two
sided", 0.95)
t-test statistic value = -0.7436128
[1] "Two-sided alternative"
p-value = 0.4649789
Alpha = 0.05
[1] "Decision: Since p > alpha, failed to reject Null hypothesis"
>
>
```

■

Table 5.6.2 gives the summary of tests for means.

5.7 OPTIMAL TEST OF HYPOTHESES

Let us consider the random variables $X_1, X_2, \ldots, X_n$ having the distribution function $F(x; \theta)$, where $\theta \in \Omega$. We consider two partitions of Ω as Θ_0 and Θ_1. Let $\theta \in \Theta_0$ or $\theta \in \Theta_1$ such that $\Theta_0 \cup \Theta_1 = \Omega$. We consider the null hypothesis as

$$H_0 : \theta \in \Theta_0 \text{ against } H_A : \theta \in \Theta_1. \tag{5.7.1}$$

Table 5.6.2 Tests for Means.

H_0	Assumptions	Test Statistic	H_A	Critical Region	p-value
$\mu = \mu_0$	Population distribution is normal, σ is known, sample size is large	$Z = \dfrac{\bar{X} - \mu_0}{\sigma/\sqrt{n}}$.	$\mu < \mu_0$ $\mu > \mu_0$	$Z < -Z_\alpha$ $Z > Z_\alpha$	$p = P(Z < -z)$ $p = P(Z > z)$
			$\mu \neq \mu_0$	$Z < -Z_{\alpha/2}$ or $Z > Z_{\alpha/2}$	$p = 2P(Z > |z|)$
$\mu = \mu_0$	Population distribution is normal, σ is unknown, sample size is small	$t = \dfrac{\bar{X} - \mu_0}{S/\sqrt{n}}$, $\nu = n - 1$.	$\mu < \mu_0$ $\mu > \mu_0$	$t < -t_\alpha$ $t > t_\alpha$	$p = P(T < -t)$ $p = P(T > t)$
			$\mu \neq \mu_0$	$t < -t_{\alpha/2}$ or $t > t_{\alpha/2}$	$p = 2P(T > |t|)$
$\mu_1 - \mu_2 = \Delta$	Population distribution is normal, σ_1 and σ_2 known	$Z = \dfrac{\bar{X}_1 - \bar{X}_2 - \Delta}{\sqrt{\sigma_1^2/n_1 + \sigma_2^2/n_2}}$.	$\mu_1 - \mu_2 < \Delta$ $\mu_1 - \mu_2 > \Delta$ $\mu_1 - \mu_2 \neq \Delta$	$Z < -Z_\alpha$ $Z > Z_\alpha$	$p = P(Z < -z)$ $p = P(Z > z)$
				$Z < -Z_{\alpha/2}$ or $Z > Z_{\alpha/2}$	$p = 2P(Z > |z|)$
$\mu_1 - \mu_2 = \Delta$	Population distribution is normal, $\sigma_1 = \sigma_2$ but unknown, sample sizes are small, samples are independent	$t = \dfrac{\bar{X}_1 - \bar{X}_2 - \Delta}{S_p\sqrt{1/n_1 + 1/n_2}}$, where $S_p = \dfrac{A+B}{n_1+n_2-2}$, $A = (n_1 - 1)S_1^2$, $B = (n_2 - 1)S_2^2$, and $\nu = n_1 + n_2 - 2$.	$\mu_1 - \mu_2 < \Delta$ $\mu_1 - \mu_2 > \Delta$ $\mu_1 - \mu_2 \neq \Delta$	$t < -t_\alpha$ $t > t_\alpha$ $t < -t_{\alpha/2}$ or $t > t_{\alpha/2}$	$p = P(T < -t)$ $p = P(T > t)$ $p = 2P(T > |t|)$
$\mu_1 - \mu_2 = \Delta$	Population distribution is normal, $\sigma_1 \neq \sigma_2$ but unknown, sample sizes are small, samples are independent	$t = \dfrac{\bar{X}_1 - \bar{X}_2 - \Delta}{\sqrt{S_1^2/n_1 + S_2^2/n_2}}$, $\nu = \dfrac{(C+D)^2}{\frac{C^2}{n_1-1} + \frac{D^2}{n_2-1}}$, where, $C = S_1^2/n_1$, $D = S_2^2/n_2$.	$\mu_1 - \mu_2 < \Delta$ $\mu_1 - \mu_2 > \Delta$ $\mu_1 - \mu_2 \neq \Delta$	$t < -t_\alpha$ $t > t_\alpha$ $t < -t_{\alpha/2}$ or $t > t_{\alpha/2}$	$p = P(T < -t)$ $p = P(T > t)$ $p = 2P(T > |t|)$
$\mu_D = \Delta$	Population distribution is normal, σ is unknown, sample sizes are small, samples are dependent	$t = \dfrac{\bar{d} - \Delta}{s_d/\sqrt{n}}$, $\nu = n - 1$.	$\mu_D < \Delta$ $\mu_D > \Delta$ $\mu_D \neq \Delta$	$t < -t_\alpha$ $t > t_\alpha$ $t < -t_{\alpha/2}$ or $t > t_{\alpha/2}$	$p = P(T < -t)$ $p = P(T > t)$ $p = 2P(T > |t|)$

We test the null hypothesis regarding the particular value of θ based on the random sample denoted by $\mathbf{X} = X_1, X_2, \ldots, X_n$ from the distribution function $F(x; \theta)$. The values of the random sample $\mathbf{X} = X_1, X_2, \ldots, X_n$ are denoted by $\mathbf{x} = x_1, x_2, \ldots, x_n$. Thus $\mathbf{x}$ will determine whether the null hypothesis is accepted or rejected. For making such a decision, we partitioned the sample space associated with $\mathbf{X}$ into two measurable sets, say C and C'. If $\mathbf{x} \in C$, we accept H_0 and if $\mathbf{x} \in C'$, we reject H_0. Set C is called the *acceptance region*, whereas set C', which is a complement of C, is called a *rejection region*. The acceptance of H_0 does not mean that H_0 will always be accepted for the given distribution; it means only for the given sample values, H_0 is a plausible hypothesis. Similarly, rejection of H_0 does not mean that H_0 will always be rejected for the given distribution; it just means that H_0 does not look plausible on the basis of the given set of observations.

Thus, a test for H_0 is a rule for rejecting or accepting the H_0 on the basis of whether the given set of observations $\mathbf{x}$ belongs or does not belong to a specified set of a critical region. Thus, for each test of H_0, there is a corresponding unique critical region, and conversely for each critical region, there corresponds a unique test. Therefore, we may say that a test is defined by its critical regions, and conversely, a critical region defines a test.

When a test procedure is applied, there are errors that we are liable to commit. As discussed in Section 5.6, these errors are known as type I errors and type II errors. A type I error is committed when H_0 is rejected though in fact it is true. A type II error is committed when H_0 is accepted though in fact it is false. One should make the choice of test for H_0 while keeping in mind these errors one may commit while using any test.

Thus, a good test should keep both types of errors in control. Since committing any type of error is a random event, in a good test we therefore try to minimize the probability of committing both types of errors. These probabilities depend on the parameter θ, and therefore the probability of type I error is defined as the probability that the $\mathbf{X}$ lies in C; that is, $P_\theta(\mathbf{X} \in C), \theta \in \theta_0$.

The size or significance level of the test is the probability of type I error, which is represented by α. Thus,

$$\alpha = P_\theta(\mathbf{X} \in C), \quad \theta \in \theta_0. \tag{5.7.2}$$

The probability of committing a type II error is the probability that $\mathbf{X}$ lies in C'.

This probability is denoted by β and is given by

$$\beta = P_\theta(\mathbf{X} \in C'), \quad \theta \in \Theta - \theta_0. \tag{5.7.3}$$

Both types of errors are serious in nature, but due to the relationship between the two errors, we cannot minimize both errors simultaneously for a fixed sample size. Thus, a test that minimizes the probabilities of type I errors, in fact, maximizes the type II errors. Thus, we keep the probability of a type I error fixed at a desirable level and try to minimize the probability of a type II error. Therefore, for a fixed size α, we select tests that minimize the type II error or equivalently maximize the probability of rejecting H_0 when $\theta \in \Theta - \theta_0$, which is also known as the *power* of the test.

The power of the test is defined as the probability of rejecting H_0 when in fact it is false. This is a measure of assurance provided by the test against accepting a false hypothesis. The power function of the test is given by

$$\gamma(\theta) = P_\theta(\mathbf{X} \in C), \quad \theta \in \Theta - \theta_0. \tag{5.7.4}$$

In selecting a test, we choose α very close to zero, and for $\theta \in \Theta - \theta_0$, $\gamma(\theta)$ should be as large as possible, close to 1.

We consider the case of testing a simple hypothesis against a simple alternative hypothesis. In this case we have Θ_0, which consists of a single element θ_0, and Θ_1, which consists of a single element θ_1. Thus, $\Omega = \{\theta_0, \theta_1\}$. We state the null hypothesis as

$$H_0 : \theta = \theta_0$$

and the alternative as

$$H_A : \theta = \theta_1.$$

We fix the significance level as α. Then the best test, which is the most powerful test, is defined in terms of best critical region because a best test will have the best critical region. The best critical region is defined next.

Definition 5.7.1. *A critical region C is called the* most powerful critical region *(best critical region) of size α, for testing $H_0 : \theta = \theta_0$ against $H_A : \theta = \theta_1$ if*

$$P_{\theta_0}(\mathbf{X} \in C) = \alpha, \tag{5.7.5}$$

and

$$P_{\theta_1}(\mathbf{X} \in C) \geq P_{\theta_1}(\mathbf{X} \in W), \tag{5.7.6}$$

for any other critical region W of the sample space satisfying

$$P_{\theta_0}(\mathbf{X} \in W) = \alpha. \tag{5.7.7}$$

Definition 5.7.1 can be extended to define the most powerful critical region for testing a simple null hypothesis against a composite alternative hypothesis.

Definition 5.7.2. *A critical region C is called the uniformly most powerful critical region (UMP critical region) of size α, for testing $H_0 : \theta = \theta_0$ against $H_A : \theta \neq \theta_0$ if*

$$P_{\theta_0}(\mathbf{X} \in C) = \alpha, \tag{5.7.8}$$

and

$$P_\theta(\mathbf{X} \in C) \geq P_\theta(\mathbf{X} \in W), \quad \text{for all } \theta \neq \theta_0 \qquad (5.7.9)$$

for any other critical region W of the sample space satisfying

$$P_{\theta_0}(\mathbf{X} \in W) = \alpha. \qquad (5.7.10)$$

In many cases, it has been observed that no uniformly most powerful critical region exists. In such cases, we need some other desirable properties of the most powerful tests so that we can decide which test is better among all the powerful tests despite having the same level of significance and power. The desirable property is called unbiasedness, which is defined next.

Definition 5.7.3. *A critical region C is said to be an* unbiased critical region *if*

$$P_\theta(\mathbf{X} \in C) \geq P_{\theta_0}(\mathbf{X} \in C) = \alpha, \quad \text{for all } \theta \neq \theta_0. \qquad (5.7.11)$$

For a biased region, we find that the probability of accepting the null hypothesis when it is false will be greater than the probability of accepting the null hypothesis when it is true. Thus, biased regions lead to undesirable property that we should avoid, and they can be avoided by selecting unbiased critical regions among all the critical regions.

Definition 5.7.4. *A critical region C is called* uniformly most powerful *among* unbiased critical region *(UMPU critical region) of size α, for testing $H_0 : \theta = \theta_0$ against $H_A : \theta \neq \theta_0$ if*

$$P_{\theta_0}(\mathbf{X} \in C) = \alpha \qquad (5.7.12)$$

$$P_\theta(\mathbf{X} \in C) \geq P_{\theta_0}(\mathbf{X} \in C) = \alpha, \quad \text{for all } \theta \neq \theta_0 \qquad (5.7.13)$$

and

$$P_\theta(\mathbf{X} \in C) \geq P_\theta(\mathbf{X} \in W), \quad \text{for all } \theta \neq \theta_0 \qquad (5.7.14)$$

for any other critical region W of the sample space satisfying

$$P_{\theta_0}(\mathbf{X} \in W) = \alpha. \qquad (5.7.15)$$

To construct the optimal critical regions, we use the lemma provided by Neyman and Pearson, which yields necessary and sufficient conditions for optimum conditions in terms of sample points.

Theorem 5.7.1 (Neyman–Pearson Theorem). *Let $X_1, X_2, \ldots, X_n$ be a random sample of fixed size n having the distribution function $F(x; \theta)$, where $\theta \in \Omega$.*

We consider two partitions of Ω as θ_0 and θ_1. Let $\theta \in \theta_0$ or $\theta \in \theta_1$ such that $\theta_0 \cup \theta_1 = \Omega$. We consider the null hypothesis as

$$H_0 : \theta = \theta_0 \quad against \quad H_A : \theta \in \theta_1.$$

We define the likelihood function of $X_1, X_2, \ldots, X_n$ as

$$L(\mathbf{x}, \theta) = \prod_{i=1}^{n} f(x_i, \theta). \tag{5.7.16}$$

Let C be a subset of the sample space such that

$$\frac{L(\mathbf{x}, \theta_0)}{L(\mathbf{x}, \theta_1)} \leq k; \quad for\ all\ \mathbf{x} \in \mathbf{C} \tag{5.7.17}$$

$$\frac{L(\mathbf{x}, \theta_0)}{L(\mathbf{x}, \theta_1)} > k; \quad for\ all\ \mathbf{x} \in \mathbf{C'} \tag{5.7.18}$$

where k is a positive number. Then C is called the most powerful critical region for testing the null hypothesis against the alternative hypothesis.

■ Example 1

Let $\mathbf{X} = X_1, X_2, \ldots, X_n$ denote a random sample from a normal distribution with PDF $N(\theta, \sigma^2)$ where σ^2 is unknown.

$$f(x; \theta) = \frac{1}{\sigma \sqrt{2\pi}} e^{-\frac{1}{2}\left(\frac{x-\theta}{\sigma}\right)^2}, \quad -\infty < x < \infty. \tag{5.7.19}$$

Find the best critical region for testing a simple null hypothesis $H_0 : \theta = \theta_1$ against the simple alternative hypothesis $H_0 : \theta = \theta_2$.

Solution

We have

$$L(\theta; \mathbf{x}) = \frac{1}{(\sigma \sqrt{2\pi})^n} e^{-\frac{1}{2\sigma^2}\left[\sum_{i=1}^{n}(x_i-\theta)^2\right]}. \tag{5.7.20}$$

Now we calculate

$$\frac{L(\theta_1; \mathbf{x})}{L(\theta_2; \mathbf{x})} = \frac{\frac{1}{(\sigma\sqrt{2\pi})^n} e^{-\frac{1}{2\sigma^2}\left[\sum_{i=1}^{n}(x_i-\theta_1)^2\right]}}{\frac{1}{(\sigma\sqrt{2\pi})^n} e^{-\frac{1}{2\sigma^2}\left[\sum_{i=1}^{n}(x_i-\theta_2)^2\right]}} \tag{5.7.21}$$

$$= e^{-\frac{1}{2\sigma^2}\left[\sum_{i=1}^{n}(x_i-\theta_1)^2 - \sum_{i=1}^{n}(x_i-\theta_2)^2\right]}. \tag{5.7.22}$$

According the Neyman–Pearson lemma, if k is a positive number and C is a subset of the sample space such that

$$\frac{L(\mathbf{x}, \theta_0)}{L(\mathbf{x}, \theta_1)} \leq k; \quad \text{for all } \mathbf{x} \in C, \tag{5.7.23}$$

then C is the best critical region.

Thus,

$$e^{-\frac{1}{2\sigma^2}\left[\sum_{i=1}^{n}(x_i-\theta_1)^2 - \sum_{i=1}^{n}(x_i-\theta_2)^2\right]} \leq k. \tag{5.7.24}$$

In this case, the inequality holds if and only if

$$\bar{x}(\theta_1 - \theta_2) \leq k' \tag{5.7.25}$$

where k' is such that the test has the significance level α.

The best critical region is the set C given by

$$C = \{(x_1, x_2, \ldots, x_n) : \bar{x} > k_1\} \quad \text{where} \quad k_1 = k'/(\theta_1 - \theta_2). \tag{5.7.26}$$

The k_1 is a constant that can be determined so that the size of the critical region is a desired significance level α. ∎

5.8 LIKELIHOOD RATIO TEST

The Neyman-Pearson lemma provides the best critical region for testing simple hypotheses against simple alternative hypotheses and can be extended to find the best critical region for testing simple null hypotheses against composite hypotheses. The Neyman-Pearson lemma bases the criteria on the magnitude of the ratio of two probability density functions obtained under simple and alternative hypothesis. The best critical region depends on the nature of the population distribution and the form of the alternative hypothesis under consideration. The likelihood ratio test procedure is based on Neyman–Pearson criteria. The likelihood ratio test procedure is defined as follows. Let X be a random variable with probability density function $f(x, \theta)$. Generally, the distribution form of the population is known and the parameter θ is unknown, and $\theta \in \Omega$, where Ω is the set of all possible values of θ, called the parameter space. The family of distribution can be constructed for $f(x, \theta)$ as $\{f(x, \theta), \theta \in \Omega\}$. For example, in case of a normal probability density function $N(\mu, \sigma^2)$, where σ^2 is known, the parameter space is defined as

$$\Omega = \{\mu : -\infty < \mu < \infty\}. \tag{5.8.1}$$

The general family of distribution is defined as

$$\{f(x; \theta_1, \theta_2, \ldots, \theta_n) : \theta_i \in \Omega, i = 1, 2, \ldots, n\}. \qquad (5.8.2)$$

The particular set of parameters called the subspace of Ω will be defined by the null hypothesis.

Let $X_1, X_2, \ldots, X_n$ be the random variable having the distribution function $F(x; \theta_1, \theta_2, \ldots, \theta_n) : \theta_i \in \Omega, i = 1, 2, \ldots, n$. We consider two partitions of Ω as ω and $\Omega - \omega$.

Let $\theta_1, \theta_2, \ldots, \theta_n \in \omega$ or $\theta_1, \theta_2, \ldots, \theta_n \in \Omega - \omega$ such that $\omega \cup \Omega - \omega = \Omega$.

We consider the null hypothesis as

$$H_0 : \theta_1, \theta_2, \ldots, \theta_n \in \omega, \qquad (5.8.3)$$

against

$$H_A : \theta_1, \theta_2, \ldots, \theta_n \in \Omega - \omega. \qquad (5.8.4)$$

The likelihood function of the random sample $x_1, x_2, \ldots, x_n$ is given by

$$L = \prod_{i=1}^{n} f(x_i; \theta_1, \theta_2, \ldots, \theta_k). \qquad (5.8.5)$$

Now we can find the likelihood estimates by using the least square principle, which is given by

$$\frac{\partial L}{\partial \theta_i} = 0, \quad i = 1, 2, \ldots, k. \qquad (5.8.6)$$

These likelihood estimates are then substituted in the likelihood function to obtain the maximum likelihood estimates under the null hypothesis and alternative hypothesis. The criterion for the maximum likelihood test λ is defined as

$$\lambda = \frac{L(x, \hat{\theta}|H_0)}{L(x, \hat{\theta}|H_1)}. \qquad (5.8.7)$$

The λ is a function of sample observations and, therefore, it is a random variable.

Since $\omega \in \Omega$, $L(x, \hat{\theta}|H_0) \leq L(x, \hat{\theta}|H_1)$ and therefore $\lambda \leq 1$. Since $\lambda \geq 0; 0 \leq \lambda \leq 1$.

The use likelihood ratio test is not fully made in bioinformatics, but some work has been started in this direction (Wang and Ethier, 2004; Bokka and

Mathur, 2006; Paik et al., 2006; Hu, 2008). Readers are encouraged to read related research papers.

To determine whether gene i is differentially expressed between control and experimental samples, Wang and Ethier (2004) used the following setup.

■ Example 1

y_{ijc} is the observed expression intensity of gene i in array j ($i = 1, \ldots, n; j = 1, \ldots, d_1$) for the control sample. y_{ije} is the observed expression intensity of gene i in array j ($i = 1, \ldots, n; j = 1, \ldots, d_2$) for the experimental sample.

The two-component model is $y_{ijc} = \alpha_c + \mu_{ic}e^{\eta_{ijc}} + \varepsilon_{ijc}$ for the control sample, and $y_{ije} = \alpha_e + \mu_{ie}e^{\eta_{ije}} + \varepsilon_{ije}$ for the experimental sample, where $\eta_{ijc} \sim N(0, \sigma_{\eta c}^2), \varepsilon_{ijc} \sim N(0, \sigma_{\varepsilon c}^2), \eta_{ije} \sim N(0, \sigma_{\eta e}^2), \varepsilon_{ije} \sim N(0, 0_{\varepsilon e}^2)$.

The likelihood ratio test statistic derived by Wang and Ethier (2004) is given by

$$
\lambda_i = \max_{\mu_{ic}=\mu_{ie}>0} \left\{ \begin{array}{l} \prod\limits_{j=1}^{d_1} \int\limits_{-\infty}^{\infty} \exp\left[\frac{-(x-\log u_{ic})^2}{2\sigma_{\eta c}^2} - \frac{(y_{ijc}-e^x)^2}{2\sigma_{\varepsilon c}^2} \right] dx \\ \times \prod\limits_{j=1}^{d_2} \int\limits_{-\infty}^{\infty} \exp\left[\frac{-(x-\log u_{ie})^2}{2\sigma_{\eta e}^2} - \frac{(y_{ije}-e^x)^2}{2\sigma_{\varepsilon e}^2} \right] dx \end{array} \right\} \Bigg/
$$

$$
\max_{\mu_{ic}>0,\mu_{ie}>0} \left\{ \begin{array}{l} \prod\limits_{j=1}^{d_1} \int\limits_{-\infty}^{\infty} \exp\left[\frac{-(x-\log u_{ic})^2}{2\sigma_{\eta c}^2} - \frac{(y_{ijc}-e^x)^2}{2\sigma_{\varepsilon c}^2} \right] dx \\ \times \prod\limits_{j=1}^{d_2} \int\limits_{-\infty}^{\infty} \exp\left[\frac{-(x-\log u_{ie})^2}{2\sigma_{\eta e}^2} - \frac{(y_{ije}-e^x)^2}{2\sigma_{\varepsilon e}^2} \right] dx \end{array} \right\}.
$$

(5.8.8)

Wang and Ethier (2004) used the data set comparing gene expression between IHF + and IHF − strains of bacterial *E. coli* (Arfin et al., 2000), which contained gene expression data for 1,973 genes, and 4 replicates are performed for both IHF + and IHF − strains. The genome-wide significance level at $\alpha = 0.1$, and the Bonferroni method (see Chapter 11) is used as the correction method. The likelihood ratio test (5.8.8) detects 88 genes, while the t-test detects 10 genes. More discussion about multiple testing procedures is presented in Chapter 11. The nonparametric likelihood ratio test is presented in Chapter 6.

Nonparametric Statistics

Most of the statistical tests discussed so far assume that the distribution of the parent population is known. Almost all the exact tests (small sample tests) are based on the assumption that the population under consideration has a normal distribution. A problem occurs when the distribution of the population is unknown. For example, gene expression data are not normal, and socioeconomic data, in general, are not normal. Similarly, the data in psychometrics, sociology, and educational statistics are seldom distributed as normal. In these cases, in which the parent population distribution is unknown, we apply nonparametric procedures. These nonparametric procedures make fewer and much weaker assumptions than those associated with the parametric tests.

The advantages of nonparametric procedures over parametric procedures are that the nonparametric procedures are readily comprehensible, very simple, and easy to apply. No assumption is made about the distribution of the parent population. If the data are mere classification, there is no parametric test procedure available to deal with the problem, but there are several nonparametric procedures available. If the data are given only in terms of ranks, then only nonparametric procedures can be applied. But if all the assumptions for the parametric procedures are satisfied, then the parametric procedures are more powerful than the nonparametric procedures.

Since in microarray data it is unreasonable to make the assumption that the parent population is normal, we find that the nonparametric methods are very useful in analyzing the microarray data.

6.1 CHI-SQUARE GOODNESS-OF-FIT TEST

The Chi-square goodness-of-fit test is the oldest distribution-free test available in the literature. It is called a distribution-free test because the distribution of the test statistic itself does not depend on the distribution of the parent population. Sometimes it is also called a nonparametric test, but it is still a distribution-free test.

The Chi-square goodness-of-fit test is used to test the hypothesis about whether or not the given data belong to a specified distribution. In other words, this test determines whether the unknown distribution from which the sample is drawn is the same as the specified distribution.

Let a random sample of size n be drawn from a population with unknown cumulative distribution function (CDF) $F_x(x)$. We would like to test the null hypothesis

$$H_0 : F_x(x) = F_0(x),$$

against the alternative hypothesis

$$H_1 : F_x(x) \neq F_0(x),$$

where $F_0(x)$ is completely specified.

The Chi-square test for the goodness-of-fit test needs the observed observations classified in categories. In n trials of a single coin, these categories are the number of heads or number of tails. Similarly, in the case of a throw of n dice, these categories are the numbers on the dice. From the distribution specified in the null hypothesis, $F_0(x)$, the probability that a random observation will be classified into each of the chosen or fixed categories is calculated. From the probability that a random variable will fall in a particular category, the expected frequency of a specified category is calculated. Under the null hypothesis, the expected frequency should be close to the observed frequency for each category under the null hypothesis.

Let the random sample of size n be grouped into k mutually exclusive categories. Let the observed frequency for the ith class, $i = 1, 2, \ldots, k$, be denoted by o_i, and the expected frequency be denoted by e_i. The Pearson's goodness-of-fit test is given by

$$W = \sum_{i=1}^{k} \frac{(o_i - e_i)^2}{e_i}. \qquad (6.1.1)$$

Under the null hypothesis, we expect the value of the W to be near to zero. For a large n, the asymptotic distribution of the W is χ^2 with $(k-1)$ degrees of freedom.

The distribution of W is approximately χ^2 under the condition that the expected frequency is not less than 5. If the expected frequency falls below 5, then the groups need to be combined so that the expected frequency remains above 5.

■ Example 1

Let the expression level of four genes fall in a unit interval $\{x : 0 < x < 1\}$. Let there be the following partitions of the interval $A_1 = \{x : 0 < x < \frac{1}{4}\}$, $A_2 = \{x : \frac{1}{4} < x < \frac{1}{2}\}$, $A_3 = \{x : \frac{1}{2} < x < \frac{3}{4}\}$, and $A_4 = \{x : \frac{3}{4} < x < 1\}$. Let the probabilities that the expression level will fall in these partitions be given by the PDF

$$f(x) = \begin{cases} 3x^2, & 0 < x < 1 \\ 0, & \text{otherwise.} \end{cases} \qquad (6.1.2)$$

Thus,

$$p_1 = P(x \in A_1) = \int_{0}^{1/4} f(x)dx = \frac{1}{64}, \; p_2 = P(x \in A_2) = \frac{7}{64},$$

$$p_3 = P(x \in A_3) = \frac{19}{64}, \text{ and } p_4 = P(x \in A_4) = \frac{37}{64}.$$

A sample of size 64 was taken, and for classes A_1, A_2, A_3, and A_4, frequencies observed were 1, 5, 20, and 38, respectively. The observed and expected frequencies are given in Table 6.1.1. Test whether the genes follow the given distribution. Use a 5% level of significance.

Solution

We have

$$F_0(x) = \int_{0}^{x} 3x^2 dx = x^3, \; 0 < x < 1.$$

Table 6.1.1 Observed and Expected Frequencies

Class	Observed (o_i)	Expected ($e_i = np_i$)
A_1	1	1
A_2	5	7
A_3	20	19
A_4	38	37

The null hypothesis is given by

$$H_0 : F_X(x) = F_0(x)$$

against the alternative hypothesis

$$H_1 : F_X(x) \neq F_0(x).$$

We make the table of expected and observed frequencies for the classes $A_1, A_2, A_3,$ and A_4.

Since the expected frequency for class A_1 is less than 5, we pool classes A_1 and A_2. The pooled frequency is as follows:

Class	Observed (o_i)	Expected ($e_i = np_i$)
A_1 and A_2	6	8
A_3	20	19
A_4	38	37

The Pearson's goodness-of-fit test statistic is given by

$$W = \sum_{i=1}^{k} \frac{(o_i - e_i)^2}{e_i} = 0.5796.$$

The critical value of the test statistic at a 5% level of significance is 2.366.

Thus, we accept the null hypothesis at a 5% level of significance. Therefore, the genes follow the given distribution (6.1.2).

The R-code for example 1 is as follows:

```
> #Example 1
>
> # Chi-Square test
>
> chisq.test(c(6,20,38), p = c(8/64, 19/64, 37/64))
        Chi-squared test for given probabilities
data: c(6, 20, 38)
X-squared = 0.5797, df = 2, p-value = 0.7484
```

■

6.2 KOLMOGOROV-SMIRNOV ONE-SAMPLE STATISTIC

In the Chi-square goodness-of-fit test, k categories are made by using n observations. The main concern about the Chi-square goodness-of-fit test is that for n observations, there are only $k \leq n$ groups. Moreover, the expected frequency for each group must be greater than 5 to get the approximate Chi-square distribution for the test statistic. Some of the tests available in the literature are based on the empirical distribution function. The empirical distribution function is an estimate of the population distribution function and is defined as the proportion of sample observations that are less than or equal to x for all real numbers x.

Let the hypothesized cumulative distribution function be denoted by $F_0(x)$ and the empirical distribution function be denoted by $S_n(x)$ for all x. If the null hypothesis is true, then the difference between $S_n(x)$ and $F_0(x)$ must be close to zero.

Thus, for a large n, the test statistic

$$D_n = \sup_x |S_n(x) - F_x(x)|, \tag{6.2.1}$$

will have a value close to zero under the null hypothesis.

The test statistic, D_n, called the *Kolmogorov-Smirnov one-sample statistic*, does not depend on the population distribution function if the distribution function is continuous; therefore, D_n is a distribution-free test statistic.

Here we define order statistic $X_{(0)} = -\infty$ and $X_{(n+1)} = \infty$, and

$$S_n(x) = \frac{i}{n}, \quad \text{for } X_{(i)} \leq x \leq X_{(i+1)}, i = 0, 1, \ldots, n. \tag{6.2.2}$$

The probability distribution of the test statistic does not depend on the distribution function $F_x(X)$ for a continuous distribution function $F_x(X)$. The asymptotic distribution of the test statistic D_n is a Chi-square.

The exact sampling distribution of the Kolmogorov-Smirnov test statistic is known, while the distribution of the Chi-square goodness-of-fit test statistic is approximately Chi-square for finite n. Moreover, the Chi-square goodness-of-fit test requires that the expected number of observations in a cell must be greater than 5, while the Kolmogorov test statistic does not require this condition. On the other hand, the asymptotic distribution of the Chi-square goodness-of-fit test statistic does not require that the distribution of the population must be continuous, but the exact distribution of the Kolmogorov-Smirnov test statistic does require that $F_x(X)$ must be a continuous distribution. The power of the Chi-square distribution depends on the number of classes or groups made.

■ Example 1

Use the Kolmogorov-Smirnov one-sample test to find whether the set of given random numbers is drawn from normal distribution.

We attempt this problem using the following R-code;

```
> #Example 1
>
> # Kolmogorov-Smirnov Test
>
> # n = number of random numbers from t-distribution
> # df = degrees of freedom
>
> n <- 1000
> df<-10
> x <- rt(n, df)
> ks.test(x, pnorm)
        One-sample Kolmogorov-Smirnov test
data: x
D = 0.0258, p-value = 0.5167
alternative hypothesis: two-sided
>
```

According to calculations, the p-value is very high, which implies that two distributions, i.e., t and normal, are the same. Thus, the test is not very powerful if distributions are not too different.　■

6.3 SIGN TEST

Let us consider a random sample $X_1, X_2, \ldots, X_n$ drawn from a population having a continuous CDF $F_x(X)$ with unknown median M. We test the hypothesis that the unknown M has some specified value, say M_0. If M_0 is a true median of $F_x(X)$, then we must have $P(X > M) = P(X < M) = 0.50$. Thus, exactly half of the total observations will lie above the median, while the remaining half will lie below the median. Now we can formally state the null and alternative hypotheses. The null hypothesis is

$$H_0 : M = M_0,$$

$$H_A : M \neq M_0.$$

Let K be the number of observations above the hypothesized median. The distribution of K is binomial with parameters n and p, with $p = 0.5$ under the null hypothesis. It is easy to note that the difference $X_i - M_0, i = 1, 2, \ldots, n$ will be either positive or negative. Thus, the distribution of K is binomial, and therefore, the rejection region can be constructed on the basis of the alternative hypothesis. We reject the null hypothesis against the right-side alternative, if $K \in R$ for $K \geq K_\alpha$, where K_α is chosen to be the smallest integer which satisfies

$$\sum_{k=k\alpha}^{N} \binom{N}{k} 0.5^N \leq \alpha. \qquad (6.3.1)$$

The sign test for a single sample can be extended to paired samples by taking the difference between two samples and considering these differences as a single sample.

The problem with the sign test is that it does not take into consideration the magnitude of differences. Hence, the contribution of the weight of a difference does not play any role in the test statistic. Moreover, the asymptotic relative efficiency with respect to its parametric competitor t-test for a single mean is very low. The improvement over the sign test is made in the form of signed rank test.

■ Example 1

Use Golub et al. (1999) data and find whether the median expression level of class AML 38 is 650.

We can use R-code to find the answer to this problem:

```
> # Example 1
>
> # Sign Test
> # Golub et al. (1999) data
> # used expression data under AML38
>
> alldata9<-read.delim("C:\Users\\happy\\Documents\\home computer-oct-4-
    2008\\Book-SKM-2\\Chapter5-Inference\\golub9-tab.txt", header = TRUE,
    sep ="\t",
na.strings = "NA")
>
> attach(alldata9)
> x<-AML.38
>
> mu<-650 # hypothesized value of median
> z <-sort(x-mu)
> n <-length(z)
> print(b <- sum(z > mu))
[1] 911
> binom.test(b,n)
        Exact binomial test
data: b and n
```

number of successes = 911, number of trials = 7129, p-value < 2.2e-16
alternative hypothesis: true probability of success is not equal to 0.5
95 percent confidence interval:
0.1201213 0.1357603
sample estimates:
probability of success
    0.1277879
Since the p-value is close to zero, we reject the claim that the hypothesized median is 650. ∎

6.4 WILCOXON SIGNED-RANK TEST

The Wilcoxon signed-rank test requires that the parent population be symmetric. Let us consider a random sample $X_1, X_2, \ldots, X_n$ from a continuous CDF F, which is symmetric about its median M. The null hypothesis can be stated as

$$H_0 : M = M_0.$$

The alternative hypotheses can be postulated accordingly. We notice that the differences $D_i = X_i - M_0$ are symmetrically distributed about zero, and therefore, the number of positive differences will be equal to the number of negative differences. The ranks of the differences $|D_1|, |D_2|, \ldots, |D_N|$ are denoted by $Rank(.)$. Then, the test statistic can be defined as

$$T^+ = \sum_{i=1}^{n} a_i \, \text{Rank} \, (|D_i|) \tag{6.4.1}$$

$$T^- = \sum_{i=1}^{n} (1 - a_i) \, \text{Rank} \, (|D_i|) \tag{6.4.2}$$

where

$$a_i = \begin{cases} 1, & \text{if} \quad D_i > 0, \\ 0, & \text{if} \quad D_i < 0. \end{cases} \tag{6.4.3}$$

Since the indicator variables a_i are independent and identically distributed Bernoulli variates with $P(a_i = 1) = P(a_i = 0) = 1/2$, under the null hypothesis

$$E(T^+|H_0) = \sum_{i=1}^{n} E(a_i) \, \text{Rank} \, |D_i| = \frac{n(n+1)}{4}, \tag{6.4.4}$$

and

$$\text{Var}(T^+|H_0) = \sum_{i=1}^{n} \text{var}(a_i) \left[\text{Rank} \, |D_i| \right]^2 = \frac{n(n+1)(2n+1)}{24}. \tag{6.4.5}$$

Another common representation for the test statistic T^+ is given as follows.

$$T^+ = \sum_{1 \leq 1 \leq J \leq n} \sum T_{ij} \tag{6.4.6}$$

where

$$T_{ij} = \begin{cases} 1, & \text{if } D_i + D_j > 0, \\ 0, & \text{otherwise.} \end{cases} \tag{6.4.7}$$

Similar expressions can be derived for T^-. The paired samples can be defined on the basis of the differences $X_1 - Y_1,\ X_2 - Y_2, \ldots, X_n - Y_n$ of a random sample of n pairs $(X_1, Y_1), (X_2, Y_2) \ldots, (X_n, Y_n)$. Now these differences are treated as a single sample, and the one-sample test procedure is applied. The null hypothesis to be tested will be

$$H_0 : M = M_0$$

where M_0 is the median of the differences $X_1 - Y_1, X_2 - Y_2, \ldots, X_n - Y_n$. These differences can be treated as a single sample with hypothetical median M_0. Then, the Wilcoxon signed-rank method described previously for a single sample can be applied to test the null hypothesis that the median of the differences is M_0.

The Wilcoxon test statistic is asymptotically normal for a moderately large sample size. This property is used to find the critical values using the normal distribution when the sample size is moderately large.

■ Example 1

A study investigated whether a particular drug may change the expression level of some of the genes of a cancer patient. The paired data sets, given in Table 6.4.1, for the expression levels of genes of a patient are obtained at time t_1 before giving the drug and time t_2 after giving the drug.

Let θ be the mean difference between the expression level at time t_1 and time t_2. Test the null hypothesis that

$$H_0 : \theta = 0 \quad \text{against} \quad H_A : \theta > 0.$$

Use a 5% level of significance.

Solution

The expression levels with signed ranks are shown in Table 6.4.2. We find that $T^+ = 45$. The critical value of the Wilcoxon signed-rank test at a 5% level of significance is 37.

Table 6.4.1 Expression Levels at Time t_1 and t_2

Gene	Expression Level at Time t_1	Expression Level at Time t_2
g1	2670.171	1588.39
g2	2322.195	1377.756
g3	887.6829	638.3171
g4	915.1707	518.0488
g5	19858.68	4784.585
g6	14586.05	3644.561
g7	44259.61	25297.95
g8	34081.98	18560.51
g9	2381.634	1557.293

Table 6.4.2 Expression Levels at Time t_1 and t_2 with Signed Ranks

Gene	Expression Level at Time t_1	Expression Level at Time t_2	Difference	Signed Rank
g1	2670.171	1588.39	1081.7810	5
g2	2322.195	1377.756	944.4390	4
g3	887.6829	638.3171	249.3658	1
g4	915.1707	518.0488	397.1219	2
g5	19858.68	4784.585	15074.1000	7
g6	14586.05	3644.561	10941.4900	6
g7	44259.61	25297.95	18961.6600	9
g8	34081.98	18560.51	15521.4700	8
g9	2381.634	1557.293	824.3410	3

We reject the null hypothesis at a 5% level of significance. Thus, the particular drug may change the expression level of some of the genes of a cancer patient.

The R-code for example 1 is as follows:

```
> # Example 1
>
> # Wilcoxon Signed-Rank Test
>
> t1<-c(2670.171,2322.195,887.6829,915.1707,19858.68,14586.05,44259.61,
34081.98,+2381.634)
> t2<-
c(1588.39,1377.756,638.3171,518.0488,4784.585,3644.561,25297.95,18560.51,
1557.29 3+)
>
```

```
> d<-(t1- t2)
> wilcox.test(d,mu = 0)
        Wilcoxon signed rank test
data: d
V = 45, p-value = 0.003906
alternative hypothesis: true location is not equal to 0
```

The value of a test statistic given by R-code is 45, and the p-value is 0.003906. Since the p-value is very small as compared to significance level, we reject the null hypothesis that the expression levels at time t_1 and t_2 are not the same. ∎

6.5 TWO-SAMPLE TEST

Let us consider two independent random samples $X_1, X_2, \ldots, X_m$ and $Y_1, Y_2, \ldots, Y_n$ from two populations X and Y, respectively. Let the populations X and Y have CDFs F_x and F_y, respectively. The null hypothesis we would like to test is that

$$H_0 : F_Y(x) = F_X(x) \quad \text{for all } x,$$

against

$$H_A : F_Y(X) = F_X(X + \theta) \quad \text{where } \theta \neq 0.$$

Here we assume that the populations are the same except for a possible shift in location θ. Thus, populations have the same shape and scale parameter, and the difference in the population location parameter is equal to zero under the null hypothesis. We have $m + n$ random variables, which can be arranged in $\binom{m+n}{m}$ ways. This sample pattern will provide information about the difference in the location between two populations, if any. Most of the tests available in the literature for detecting the shift in location are based on some type of function of combined arrangements of two samples.

6.5.1 Wilcoxon Rank Sum Test

The Wilcoxon rank sum test statistic is calculated by using the combined ordered samples. We order combined samples of size $N = m + n$, of X-values and Y-values from least to greatest or greatest to least value. Now we determine the ranks of the X-value in the combined sample and let the rank of X_1 be R_1, X_2 be $R_2, \ldots, X_m$ be R_m in the combined ordered sample. Then, the Wilcoxon rank sum test is given by

$$W = \sum_{i=1}^{m} R_i. \qquad (6.5.1.1)$$

To test the null hypothesis that the location parameter θ is zero, we set up the null hypothesis as

$$H_0 : \theta = 0,$$

against

$$H_A : \theta > 0.$$

We reject the null hypothesis if $W \geq w_\alpha$, where w_α is a constant chosen to make the type I error probability equal to α, the level of significance.

For the alternative hypothesis

$$H_A : \theta < 0,$$

we reject the null hypothesis if

$$W \leq n(m + n + 1) - w_\alpha. \qquad (6.5.1.2)$$

For the alternative hypothesis

$$H_A : \theta \neq 0,$$

we reject the null hypothesis if

$$W \geq w_{\alpha/2} \quad \text{or} \quad W \leq n(m + n + 1) - w_{\alpha/2}. \qquad (6.5.1.3)$$

If we use the data given in Example 1 and assume that expression levels at time t_1 before giving the drug and time t_2 are independent, then we can apply the Wilcoxon rank sum test.

The R-code is as follows:

```
> # Example 1
>
> # Wilcoxon Rank Sum Test
>
> t1<-c(2670.171,2322.195,887.6829,915.1707,19858.68,14586.05,44259.61,
34081.98,+2381.634)
> t2<-
c(1588.39,1377.756,638.3171,518.0488,4784.585,3644.561,25297.95,18560.51,
1557.29 3+)
>
> wilcox.test(t1,t2,paired = FALSE, conf. int = TRUE, conf.level = 0.95)
      Wilcoxon rank sum test
data: t1 and t2
```

W = 52, p-value = 0.3401
alternative hypothesis: true location shift is not equal to 0
95 percent confidence interval:
-2462.39 18961.66
sample estimates:
difference in location
        1112.878

Since the p-value is very large as compared to significance level, we fail to reject the null hypothesis. In example 1 in Section 6.4, we rejected the null hypothesis when we used the Wilcoxon signed-rank test. Thus, there is a huge difference in the computed value of the p-value just by a change in the assumption. This example shows the importance of the assumption of independence of expression levels obtained under two conditions.

6.5.2 Mann-Whitney Test

As we know, the sample arrangement of $m + n$ random variables provides important information about the possible difference in the shift parameters between two populations. The Mann-Whitney test is based on the concept that if two samples are arranged in a combined order, it will exhibit a particular pattern that will enable us to find whether the two populations have a difference in shift parameters. If Xs are greater than most of the Ys or vice versa, it would indicate that the null hypothesis is not true.

We combine the two samples $X_1, X_2, \ldots, X_m$ and $Y_1, Y_2, \ldots, Y_n$ and arrange them in increasing or decreasing order of magnitude. The Mann-Whitney U statistic is defined as the number of times a Y precedes an X in the combined ordered arrangement. Thus, we can define the indicator function

$$D_{ij} = \begin{cases} 1, & \text{if } Y_j < X_i, \\ 0, & \text{otherwise.} \end{cases} \qquad (6.5.2.1)$$

for $i = 1, 2, \ldots, m; j = 1, 2, \ldots, n.$

The Mann-Whitney U statistic is defined as

$$U = \sum_{i=1}^{m} \sum_{j=1}^{n} D_{ij}. \qquad (6.5.2.2)$$

Under the null hypothesis, the number of Ys greater than Xs must be the same as the number of Xs greater than Ys. Also, the statistic U/mn is a consistent statistic.

The U statistic can be expressed in terms of the Wilcoxon signed-rank test statistic:

$$U = mn + \frac{m(m+1)}{2} - T^+. \tag{6.5.2.3}$$

Thus, the Wilcoxon signed-rank test and the Mann-Whitney U test are equivalent to each other.

Under the null hypothesis, we have

$$\mu = E(U) = \frac{mn}{2}, \tag{6.5.2.4}$$

and

$$\sigma^2 = \text{Var}(U) = \frac{mn(m+n+1)}{12}. \tag{6.5.2.5}$$

We can easily show that

$$Z = \frac{U - \mu}{\sigma} \sim N(0, 1). \tag{6.5.2.6}$$

The asymptotic relative efficiency (ARE) of the Mann-Whitney U test statistic with respect to two-sample t-tests is at least 0.8664. Under the normal population, the ARE of the Mann-Whitney U test statistic with respect to two-sample t-tests is 0.955. Among all the available test statistics for two-sample location problems, the Mann-Whitney U statistic is perhaps the most powerful test statistic. There are some other test statistics available now that are a little more powerful than the Mann-Whitney U test statistic, but those test statistics require some additional assumptions.

In practice, it might not be appropriate to use the t-test for the differential expression because the expression levels are generally expected to have a normal distribution and the variances of the two types are also not expected to be equal. These two conditions are a must for applying the t-test. Thus, if the assumptions for two-sample t-tests are not satisfied, it is recommended that one should use the nonparametric tests such as the Wilcoxon-Mann-Whitney test.

■ Example 1

Consider the expression data, given in Table 6.5.2, obtained from cancer patients and healthy control subjects for gene A. Test the null hypothesis that the cancer and healthy groups have the same mean expression level for gene A, assuming that the cancer and control groups are independent groups. Use a 5% level of significance.

Table 6.5.2 Expression Levels of Gene A	
Expression Level of Gene A in Cancer Patient	Expression Level of Gene A in Healthy Control Subject
0.342	0.55
0.794	0.51
0.465	0.888
0.249	0.98
0.335	0.514
0.499	0.645
0.295	0.376
0.796	0.778
0.66	0.089

Solution

Since we do not have any information about the distributional form of the cancer patients and healthy control subjects, we cannot apply a two-sample independent t-test. We apply the Wilcoxon-Mann-Whitney test that does not require any knowledge of the distribution.

The U statistic is given by

$$U = \sum_{i=1}^{m} \sum_{j=1}^{n} D_{ij} = 27.$$

The critical value of U is 104 at a 5% level of significance, which leads to the acceptance of the null hypothesis. Thus, the cancer patients and healthy control subjects have the same expression level for gene A.

The R-code for the Mann-Whitney U test is the same as that of the Wilcoxon rank sum test:

```
> #Example 1
>
> # Mann-Whitney/Wilcoxon Rank Sum Test
>
> cancerpatient<-c(0.342,0.794,0.465,0.249,0.335,0.499,0.295,0.796,0.66)
> healthycontrol<-c(0.55,0.51,0.888,0.98,0.514,0.645,0.376,0.778,0.089)
>
> wilcox.test(cancerpatient,healthycontrol,paired = FALSE,conf.int = TRUE, conf.level = 0.95)
        Wilcoxon rank sum test
data: cancerpatient and healthycontrol
```

W = 28, p-value = 0.2973

alternative hypothesis: true location shift is not equal to 0

95 percent confidence interval:

-0.35 0.16

sample estimates:

difference in location

   -0.127

Because the p-value obtained by using the R-code is 0.2973, which is more than a 5% level of significance, we fail to reject the null hypothesis.   ∎

6.6 THE SCALE PROBLEM

In the case of a location problem, we considered the change in the shift between two populations. Under the scale problem, we are concerned about the change in the variability or dispersion between two populations. For a single sample case, the scale problem reduces to testing whether the population has a specific scale parameter. The tests designed for detecting the change in the location parameters may not be sensitive to a change in the scale parameter and hence may not be very effective in detecting the smallest change in a shift parameter or testing whether the population has a specific scale parameter. Therefore, some other tests were specifically developed for one-sample as well as two-sample scale problems.

Let us consider two independent random samples $X_1, X_2, \ldots, X_m$ and $Y_1, Y_2, \ldots, Y_n$ from two continuous populations X and Y, respectively. Let the populations X and Y have CDFs F and G, respectively, with the location parameters δ_1 and δ_2, respectively, and scale parameters η_1 and η_2, respectively. The null hypothesis is given as

$$H_0 : F(t) = G(t) \quad \text{for all } t \tag{6.6.1}$$

against

$$H_A : F\left(\frac{t - \delta_1}{\eta_1}\right) = G\left(\frac{t - \delta_2}{\eta_2}\right), \quad -\infty < t < \infty. \tag{6.6.2}$$

The most common parametric test available for testing a shift in scale parameter is called the F-test, which does not need the assumption of equality of a location parameter but needs the assumption that parent populations are normally distributed. The parameter of interest is the ratio of the scale parameter $\theta = \frac{\eta_1}{\eta_2}$. If the variance of population X, $\text{Var}(X)$, exists and both population X and Y are equal in distribution in the sense that $\frac{X - \delta}{\eta_1} \overset{d}{=} \frac{Y - \delta}{\eta_2}$ where δ is the common media of population of X and Y, then the variance of population Y, $\text{Var}(Y)$, also exists and the ratio of population variances $\theta = \left[\frac{\text{Var}(X)}{\text{Var}(Y)}\right]$ also exists. Thus, the null hypothesis mentioned in (6.5.1) reduces to

$$H_0 : \theta = 1, \qquad (6.6.3)$$

which is the same as the statement that the population scale parameters are equal.

As pointed out earlier, it is not reasonable to assume that the gene expressions are normally distributed; therefore, it will not be appropriate to use the F-test for testing the shift in scale parameter for the gene expression data. Next, we will discuss some of the nonparametric tests available in the literature for the above-mentioned hypotheses.

6.6.1 Ansari-Bardley Test

We combined two samples $X_1, X_2, \ldots, X_m$ and $Y_1, Y_2, \ldots, Y_n$. Order the combined samples from least to greatest. The score 1 is assigned to both the smallest as well as the largest observations in the combined ordered samples; similarly, score 2 is assigned to the second smallest observation and the second largest observation; and we continue in this manner. If $N = m + n$ is even, then we will have scores, $1, 2, 3, \ldots, N/2, N/2, N/2 - 1, \ldots, 3, 2, 1$. Similarly if N is odd, then we will have scores $1, 2, 3, \ldots, (N-1)/2, (N+1)/2, (N-1)/2, (N-1)/2 - 1, \ldots, 3, 2, 1$. Assume that medians of the two populations are equal. For testing the null hypothesis

$$H_0 : \theta = 1$$

we can define the Ansari-Bradley test statistic T in terms of the scores assigned to Y values, which is given by

$$T_1 = \sum_{i=1}^{n} S_i \qquad (6.6.1.1)$$

where S_i is the score assigned to Y_i.

If we consider a one-sided alternative hypothesis

$$H_A : \theta > 1,$$

then the rejection criterion is

$$\text{Reject } H_0 \text{ if } T_1 > T_\alpha,$$

at α level of significance where the constant T_α is chosen so that the probability of committing a type I error is equal to α.

If we consider a one-sided alternative hypothesis

$$H_A : \theta < 1,$$

then the rejection criterion is

$$\text{reject } H_0 \text{ if } T_1 < [T_{1-\alpha} - 1],$$

at α level of significance where the constant T_α is chosen so that the probability of committing a type I error is equal to α.

If we consider a two-sided alternative hypothesis

$$H_A : \theta \neq 1,$$

then the rejection criterion is

$$\text{reject } H_0 \text{ if } T_1 > T_{\alpha_1} \text{ or } T < [T_{1-\alpha_2} - 1],$$

at $\alpha = \alpha_1 + \alpha_2$ level of significance where the constant T_α is chosen so that the probability of committing a type I error is equal to α.

For testing the null hypothesis that $\theta = \theta_0$, where θ_0 is any specified positive constant other than 1 when the common median of population X and Y is δ_0, then we need to modify the given observations as follows. We write

$$X_i' = \frac{X - \delta_0}{\theta_0}, i = 1, 2, \ldots, m; \text{ and } Y_j' = Y - \delta_0, j = 1, 2, \ldots, n.$$

Then compute T_1 given in (6.6.1.1) using these modified observations X_i' and Y_j'.

■ Example 1

Consider the gene measurements in six cancer patients and six controls for the gene ID AC04220, given in Table 6.6.1.1:

Table 6.6.1.1 Gene Measurements in Cancer Patients and Controls

Gene ID	Patient 1	Patient 2	Patient 3	Patient 4	Patient 5	Patient 6
AC04220	10	11	7	9	17	15

Gene ID	Control 1	Control 2	Control 3	Control 4	Control 5	Control 6
AC04220	12	13	16	18	8	14

Table 6.6.1.2 Ansari-Bradley Scores

Observation	7	8	9	10	11	12	13	14	15	16	17	18
Score	1	2	3	4	5	6	6	5	4	3	2	1
Category	X	Y	X	X	X	Y	Y	Y	X	Y	X	Y

Is the variance different between control and patient? Assume that the medians of two populations are the same. Use a 10% level of significance.

Solution

Let the ratio of population variance $\theta = \left[\frac{\text{Var}(X)}{\text{Var}(Y)}\right]$ exist. We set up the null hypothesis as

$$H_0 : \theta = 1$$

against the two-sided alternative hypothesis

$$H_A : \theta \neq 1.$$

For calculating the Ansari-Bradley test statistic, we arrange the data in increasing order and assign the score 1 to the lowest and greatest observation and continue to assign the scores in this manner. Let X denote the cancer group and Y denote the control group.

Then,

$$T_1 = \sum_{i=1}^{n} S_i = 20.$$

The critical value T_{005} is approximately 28. Therefore, the null hypothesis is accepted. Thus, the variance of control and patient groups is equal.

If we apply the Ansari-Bradley test on the data provided in Example 2, we get the following results using R-code:

```
> #Example 2
>
> # Ansari-Bradley Sum Test
>
> cancerpatient<-c(0.342,0.794,0.465,0.249,0.335,0.499,0.295,0.796,0.66)
> healthycontrol<-c(0.55,0.51,0.888,0.98,0.514,0.645,0.376,0.778,0.089)
>
> ansari.test(cancerpatient,healthycontrol,paired = FALSE,conf.int = TRUE, conf.level
= 0.95)
        Ansari-Bradley test
```

```
data: cancerpatient and healthycontrol
AB = 42, p-value = 0.6645
alternative hypothesis: true ratio of scales is not equal to 1
95 percent confidence interval:
NA NA
sample estimates:
ratio of scales
      4.6464
Warning message:
In cci(alpha):
    samples differ in location: cannot compute confidence set, returning NA
>
```

We apply the Ansari-Bradley test to the data given in example 1 using R-code as follows:

```
> #Example 1
>
> # Ansari-Bradley Sum Test
>
> AC04220cancerpatient<-c(10, 11, 7, 9, 17, 15)
> AC04220healthycontrol<-c(12, 13, 16, 18, 8, 14)
>
> ansari.test(AC04220cancerpatient,AC04220healthycontrol,paired = FALSE,
conf.int = TRUE, conf.level = 0.95)
      Ansari-Bradley test
data: AC04220cancerpatient and AC04220healthycontrol
AB = 19, p-value = 0.6385
alternative hypothesis: true ratio of scales is not equal to 1
95 percent confidence interval:
  NA NA
sample estimates:
ratio of scales
      1.669643
Warning message:
In cci(alpha):
    samples differ in location: cannot compute confidence set, returning NA
>
```

6.6.2 Lepage Test

The Ansari-Bradley test is designed for situations in which the medians of two populations are equal, but the two populations may have different scale parameters. In differential gene expressions, it may be difficult to verify that the medians of two unknown populations are equal; therefore, we need to test for either location or dispersion (scale).

Let us consider two independent random samples $X_1, X_2, \ldots, X_m$ and $Y_1, Y_2, \ldots, Y_n$ from two continuous populations X and Y, respectively. Let the populations X and Y have CDFs F and G, respectively, with location parameter δ_1 and δ_2, respectively, and scale parameter η_1 and η_2, respectively. The null hypothesis is given by

$$H_0 : F(t) = G(t) \quad \text{for all } t. \tag{6.6.2.1}$$

We can also restate (6.6.2.1) as

$$H_0 : \delta_1 = \delta_2, \text{ and } \eta_1 = \eta_2, \tag{6.6.2.2}$$

against

$$H_A : \delta_1 \neq \delta_2, \text{ and/or } \eta_1 \neq \eta_2. \tag{6.6.2.3}$$

Now we calculate the Ansari-Bradley test statistic T_1, defined in (6.6.1.1). We can also calculate the Wilcoxon rank sum test statistic W defined in (6.5.1.1). The Lepage rank statistic is then defined by

$$T_L = \frac{\left[W - E(W|H_0)\right]^2}{\text{Var}(W|H_0)} + \frac{\left[T_1 - E(T_1|H_0)\right]^2}{\text{Var}(T_1|H_0)}. \tag{6.6.2.4}$$

Here $E(W|H_0)$ and $E(T_1|H_0)$ are the expected values of W and T_1 under the null hypothesis. Similarly, $\text{Var}(W|H_0)$ and $\text{Var}(T_1|H_0)$ are the variances under the null hypothesis.

Here

$$E(T_1|H_0) = \frac{n(N+2)}{4}, \tag{6.6.2.5}$$

and

$$\text{Var}(T_1|H_0) = \frac{mn(N+2)(N-2)}{48(N-1)}. \tag{6.6.2.6}$$

Also,

$$E(W|H_0) = \frac{n(N+1)}{2}, \tag{6.6.2.7}$$

$$\text{Var}(W|H_0) = \frac{mn(N+1)}{12}. \tag{6.6.2.8}$$

To test the null hypothesis given in (6.6.2.2), against (6.6.2.3), we reject the null hypothesis at an α level of significance if

$$T_L \geq t_{L,\alpha}$$

where the constant $t_{L,\alpha}$ is the critical value chosen such that the type I error probability is equal to α.

6.6.3 Kolmogorov-Smirnov Test

Let us consider two independent random samples $X_1, X_2, \ldots, X_m$ and $Y_1, Y_2, \ldots, Y_n$ from two continuous populations X and Y, respectively. Let the populations X and Y have CDFs F and G, respectively. Suppose we are interested in assessing whether the two given populations are different from each other in any respect regarding location or scale.

We formulate the null hypothesis as follows:

$$H_0 : F(t) = G(t) \quad \text{for all } t, \tag{6.6.3.1}$$

against

$$H_A : F(t) \neq G(t) \quad \text{for at least one } t. \tag{6.6.3.2}$$

The Kolmogorov-Smirnov test statistic is based on the empirical distribution functions $F_m(t)$ and $G_n(t)$ for the X and Y samples, respectively. The $F_m(t)$ and $G_n(t)$ are the estimates of the underlying distribution functions F and G, respectively, under the null hypothesis.

We define, for every real t,

$$F_m(t) = \frac{\text{number of sample values } X\text{'s} \leq t}{m}, \tag{6.6.3.3}$$

$$G_n(t) = \frac{\text{number of sample values } Y\text{'s} \leq t}{n}, \tag{6.6.3.4}$$

and

$$K = \text{greatest common divisor of } m \text{ and } n.$$

Then, the Kolmogorov-Smirnov test statistic is defined as

$$T_{KS} = \max_{(-\infty < t < \infty)} \{| F_m(t) - G_n(t)|\}. \tag{6.6.3.5}$$

We notice that the empirical distribution functions $F_m(t)$ and $G_n(t)$ are the step functions changing functional values only at the observed sample values X and Y, respectively. Thus, if we assume that $W_{(1)}, W_{(2)}, \ldots, W_{(N)}$ are the $N = m + n$ ordered values in the combined samples of $X_1, X_2, \ldots, X_m$ and $Y_1, Y_2, \ldots, Y_n$, then the Kolmogorov-Smirnov test statistic given in (6.6.3.5) can be written as

$$T_{KS} = \max_{i=1,2,\ldots,N} \{| F_m(W_{(i)}) - G_n(W_{(i)})|\}. \tag{6.6.3.6}$$

For testing (6.6.3.1) against (6.5.3.2), we reject the null hypothesis if $T_{KS} \geq T_{KS,\alpha}$; otherwise, we do not reject the null hypothesis. Here $T_{KS,\alpha}$ is a constant such that the probability of a type I error is equal to α.

■ Example 2

Consider the gene expression data, given in Table 6.6.3.1, collected for gene S from two independent groups: group A and group B. Test whether two groups have the same location and same variance, assuming that two samples X and Y are independent and drawn from two continuous populations F and G, respectively.

Solution

The greatest common divisor for $m = 6$ and $n = 6$ is 6.

We have

$$N = m + n = 12.$$

We calculate $W_{(i)}^w$'s, $F_6(W_{(i)})$, and $G_6(W_{(i)})$, which are shown in Table 6.6.3.2.

Thus, we find that

$$\max_{i=1,2,\ldots,12}\{|F_6(W_{(i)}) - G_6(W_{(i)})|\} = \frac{3}{6}.$$

Hence,

$$T_{KS} = \max_{i=1,2,\ldots,12}\{|F_6(W_{(i)}) - G_6(W_{(i)})|\} = \frac{3}{6} = 0.5.$$

The p-value for $m = n = 6$ and $T_{KS} = 6$ is $2(0.223) = 0.446$ (using a table for the critical values). Since $p = 0.223$ is more than $\alpha = 0.05$, we fail to reject the null hypothesis. Therefore, the mean and variance of two groups are the same for gene S.

Table 6.6.3.1 Expression Levels of Gene S in the Two Groups A and B	
Group A	Group B
0.765	0.625
0.165	0.064
0.743	0.580
0.355	0.342
0.671	0.936
0.688	0.605

Table 6.6.3.2 Empirical Distributions in Two Groups A and B

i	$W_{(i)}$	$F_6(W_{(i)})$	$G_6(W_{(i)})$	$\{|F_m(W_{(i)}) - G_n(W_{(i)})|\}$
1	0.064	0/6	1/6	1/6
2	0.165	1/6	1/6	0
3	0.342	1/6	2/6	1/6
4	0.355	2/6	2/6	0
5	0.58	2/6	3/6	1/6
6	0.605	2/6	4/6	2/6
7	0.625	2/6	5/6	3/6
8	0.671	3/6	5/6	2/6
9	0.688	4/6	5/6	1/6
10	0.743	5/6	5/6	0
11	0.765	6/6	5/6	1/6
12	0.936	6/6	6/6	0

There are several other procedures in the literature that are nonparametric in nature and are useful in bioinformatics.

The following R-code gives the p-value of the Kolmogorov-Smirnov test for example 2.

```
> # Example 2
> # Kolmogorov-Smirnov two sample
> groupA<-c(0.765, 0.165, 0.743, 0.355, 0.671, 0.688)
> groupB<-c(0.625, 0.064, 0.580, 0.342, 0.936, 0.605)
> ks.test(groupA, groupB, alternative = c("two.sided"),
+       exact = NULL)
    Two-sample Kolmogorov-Smirnov test
data: groupA and groupB
D = 0.5, p-value = 0.474
alternative hypothesis: two-sided
```

■

6.7 GENE SELECTION AND CLUSTERING OF TIME-COURSE OR DOSE-RESPONSE GENE EXPRESSION PROFILES

The genetic regulatory network is a system in which proteins and genes bind to each other and act as a complex input-output system for controlling cellular functions. A correct regulatory network is needed for a cell cycle to work normally. The nature and functions of pathways in the regulatory network are of great importance in the study of biological problems. It is assumed that genes with a common functional role have similar expression patterns across different experiments. The similarity may be due to the coregulations of genes in the same functional group. But this assumption does not hold for some of the gene functions, and one can find many genes that are coexpressed and not coregulated. Thus, clustering techniques (Barash and Friedman, 2002) that are based on similarity or coregulation are not valid universally. Moreover, clustering techniques and association rule mining techniques assume that the gene expression levels are measured under the same conditions or the same time points and do not take any time-lagged relationships into consideration. In the time series data, we find that most of the genes do not regulate each other simultaneously, but do so after a certain time lag. Thus, after a lag of time, the gene expression of a certain gene may affect the gene expression of another gene. Such regulation is of two kinds: activation, in which the expression level of certain genes increases due to other genes after a lag of time; and inhibition, in which the expression level of certain genes decreases because of other genes after a lag of time (Ji and Tan, 2005). Regression modeling of gene expression trajectories has been proposed as an important alternative to clustering methods for analyzing the expression patterns of cell-cycle-related genes. Liu et al. (2004) proposed a nonlinear regression model for quantitatively analyzing periodic gene expression in the studies of experimentally synchronized cells. The model accounts for the observed attenuation in cycle amplitude by a simple and biologically plausible mechanism. The expression level for each gene is represented as an average across a large number of cells. For a given cell-cycle gene, the expression in each cell in the culture is modeled as having the same sinusoidal function except that the period, which in any individual cell must be the same for all cell-cycle genes, varies randomly across cells. These random periods are modeled by using a lognormal distribution.

Most of the testing procedures for time-lagged analysis suggested in the literature are based on the pairs of genes assuming that the expression levels are measured under uniform conditions. These methods are based on the ranking of pairs of genes on the basis of some scores. The correlation method finds whether two variables have a significant linear relationship with each other using Pearson's correlation coefficient. The correlation method has been used heavily in many clustering methods. However, the correlation coefficient may be absurd sometimes, which means that it may not represent the true relationship between two variables (Peddada et al., 2003). The edge detection method (Chen et al., 1999) sums the number of edges of two gene expression curves where the edges have the same direction within a reasonable time lag to generate a score. In the edge detection method, the direction of regulation is not used for computing the score, and it can only determine potential activation relationship. Filkov et al. (2002) focused on improving the local edge detection ability. The Bayesian approach (Friedman et al., 2000; Spirtes et al., 2000) represents the structure of a gene regulatory network in a graphical way. It becomes computationally cumbersome if all aspects of gene expression data are considered. The Event Method (Kwon et al., 2003) compares two gene expression curves using events that are in the specific time interval representing the directional change of the gene expression curve at that time point, but it depends heavily on the choice of a smoothing factor.

A fractal analysis approach (Mathur et al., 2006) is used to identify hair cycle-associated genes from a time-course gene expression profile. Fractals are disordered systems with the disorder described by nonintegral dimensions, the fractal dimension. Surface reactions exhibit fractal-like kinetics (Kopelman, 1998). The kinetics have anomalous reaction orders and time-dependent coefficients (e.g., binding or dissociation). The equations involved for the binding and the dissociation phases for analyte-receptor binding are given in the following sections to permit easier reading. These equations have been applied to other analyte-receptor reactions occurring on biosensor surfaces (Butala et al., 2003; Sadana, 2003). A single fractal analysis is adequate to describe the binding and dissociation kinetics.

6.7.1 Single Fractal Analysis

Binding Rate Coefficient

Havlin (1989) indicated that the diffusion of a particle (analyte [Ag]) from a homogeneous solution to a solid surface (e.g., receptor [Ab]-coated surface) on which it reacts to form an analyte-receptor complex (Ab.Ag) is given by

$$(\text{Ab.Ag}) \approx \begin{cases} t^{(3-D_{f,\text{bind}})/2} = t^p & (t < t_c) \\ t^{1/2} & (t > t_c) \end{cases}. \qquad (6.7.1.1)$$

Here $D_{f,\text{bind}}$ or D_f is the fractal dimension of the surface during the binding step.

The t_c is the cross-over value. The binding rate coefficient, k, is the proportionality coefficient of equation (6.7.1.1). The cross-over value may be determined by $r_c^2 \sim t_c$. For a homogeneous surface where D_f is equal to 2, and when only diffusional limitations are present, $p = 1/2$, as it should be. Another way of looking at the $p = 1/2$ case (where $D_{f,\text{bind}}$ is equal to 2) is that the analyte in solution views the fractal object—in our case, the receptor-coated biosensor surface—from a "large distance." In essence, in the binding process, the diffusion of the analyte from the solution to the receptor surface creates a depletion layer of width $(Dt)^{1/2}$, where D is the diffusion constant. This gives rise to the fractal power law, $(\text{Analyte.Receptor}) \sim t^{(3-D_{f,\text{bind}})/2}$. For the present analysis, t_c is arbitrarily chosen, and we assume that the value of the t_c is not reached. One may consider the approach as an intermediate "heuristic" approach that may be used in the future to develop an autonomous (and not time-dependent) model for diffusion-controlled kinetics.

Dissociation Rate Coefficient

The diffusion of the dissociated particle (receptor [Ab] or analyte [Ag]) from the solid surface (e.g., analyte [Ag]-receptor [Ab] complex coated surface) into a solution may be given, as a first approximation, by

$$(Ab.Ag) \approx -t^{(3-D_{f,\text{diss}})/2} = t \quad (t > t_{\text{diss}}). \qquad (6.7.1.2)$$

Here $D_{f,\text{diss}}$ is the fractal dimension of the surface for the dissociation step. t_{diss} represents the start of the dissociation step. This corresponds to the highest concentration of the analyte-receptor complex on the surface. Henceforth, its concentration only decreases. The dissociation kinetics may be analyzed in a manner "similar" to the binding kinetics. The dissociation rate coefficient, k_d, is the proportionality coefficient of equation (6.7.1.2).

Lin et al. (2004) identified the hair-cycle associated genes and classified them using a computational approach. The changes in the expression of genes were observed at synchronized and asynchronized time points. Using the probabilistic clustering algorithm, Lin et al. classified the hair-cycle related genes into distinct time-course profile groups. The algorithm was used to identify the genes that are likely to play a vital role in hair-cycle morphogenesis and cycling.

In fractal analysis, after obtaining the values for the rate coefficients and the fractal dimensions for the binding and the dissociation, it is relatively simple to find out the contribution of individual genes in different phases. The method is easy to apply and provides better physical insight into gene expression profiles in each cluster. One is able to identify the cluster, which is more effectively expressed during the anagen phase of hair growth. The most significant feature of the fractal analysis is that with the fractal analysis it is possible to make quantitative the contribution made by gene expression by a single number, i.e., fractal dimension. With fractal dimension in hand, it becomes relatively simple to compare many clusters at a time, and it becomes easy to compare different replicates of an experiment, but with the computational method (Lin et al., 2004), it is not possible.

The fractal dimension is inversely proportional to the variance within the cluster. Thus, the higher the variance within the cluster, the lower the fractal dimension and vice versa. Therefore, the fractal dimension also identifies the irregular hair-growth patterns within the cluster. Thus, through the fractal dimension, we can predict the irregularity in the growth pattern in a cluster. We find that the identification of an irregular growth pattern is also one of the important issues in genetics. This type of identification of an irregular growth pattern was not achieved by Lin et al. (2004). Thus, we see that through fractal analysis, we can achieve many goals that may be useful for researchers in genetics. Fractal analysis does not require any distributional assumption for the parent population, and therefore, it is *strictly* a nonparametric method in nature.

6.7.2 Order-Restricted Inference

Understanding changes in the gene expression across a series of time points when a cell/cell line/tissue is exposed to a treatment over a time period or different dose levels is of much importance to researchers (Simmons and Peddada, 2007). For example, Perkins et al. (2006) examined *in vitro* exposed primary hepatocyte cells and *in vivo* exposed rat liver tissue to understand how these systems compared in gene expression responses to RDX. Gene expression was analyzed in primary cell cultures exposed to 7.5, 15, or 30 mg/L RDX for 24 and 48 hours. Primary cell expression effects were compared to those in liver tissue isolated from female Sprague-Dawley rats 24 hours after gavage with 12, 24, or 48 mg/Kg RDX. Samples were assessed within time point and cell type (24 hr cell, 48 hr cell, and liver tissue) to their respective controls.

The microarray experiments and dose-response studies usually have some kind of inherent ordering system. Peddada et al. (2003) took advantage of the ordering in a time-course study by using the order-restricted statistical inference, which is an efficient tool to use this ordering information. They developed a strong methodology based on order-restricted inference to analyze such gene expression data and developed user-friendly software called ORIOGEN to implement this methodology, with some modifications. ORIOGEN selects statistically significant genes and clusters genes that have similar profiles across treatment groups. The linear/quadratic regression–based method clusters genes based on statistical significance of various regression coefficients. They used known inequalities among parameters for the estimation.

Following are the steps used in the procedure:

1. Define potential candidate profiles of interest and express them in terms of inequalities between the expected gene expression levels at various time points.

2. For a given candidate profile, estimate the mean expression level of each gene using the procedure developed in Hwang and Peddada (1994).

3. The best-fitting profile for a given gene is selected using the goodness-of-fit criterion and bootstrap test procedure developed in Peddada et al. (2001).

4. A pair of genes g_1 and g_2 fall into the same cluster if all the inequalities between the expected expression levels at various time points are the same, which would imply that they follow the same temporal profile.

In a time-course experiment, let there be n time points, $T = 1, 2, \ldots, n$, and at each time point there are M arrays, each with G genes. Let Y_{igt} denote the ith expression measurement taken on gene g at time point t. Let $\overline{Y}_{gt}$ denote the sample mean of gene g at time point t and let $\overline{Y}_g = (\overline{Y}_{g1}, \overline{Y}_{g2}, \ldots, \overline{Y}_{gn})'$. The unknown true mean expression level of gene g at time point t is $E(\overline{Y}_{gt}) = \mu_{gt}$.

Inequalities between the components of $\mu_g = (\mu_{g1}, \mu_{g2}, \ldots, \mu_{gn})'$ define the true profile for gene g. Their procedure seeks to match a gene's true profile, estimated from the observed data, to one of a specified set of candidate profiles.

Some of the inequality profiles can be of the type shown here. The subscript g is dropped for the notational convenience.

Null profile: $C_0 = \{\mu \in R^T : \mu_1 = \mu_2 = \cdots = \mu_n\}$.

Monotone increasing profile (*simple order*):

$$C = \{\mu \in R^T : \mu_1 \leq \mu_2 \leq \cdots \leq \mu_n\}.$$

(with at least one strict inequality).

Monotone decreasing profile (*simple order*):

$$C = \{\mu \in R^T : \mu_1 \geq \mu_2 \geq \cdots \geq \mu_n\}.$$

Up-down profile maximum at i (*umbrella order*):

$$C = \{\mu \in R^T : \mu_1 \leq \mu_2 \leq \cdots \leq \mu_i \geq \mu_{i+1} \geq \mu_{i+2} \geq \cdots \geq \mu_n\}$$

(with at least one strict inequality among $\mu_1 \leq \mu_2 \leq \cdots \leq \mu_i$ and one among $\mu_{i+1} \geq \mu_{i+2} \geq \cdots \geq \mu_n$).

Down-up profile minimum at i (*umbrella order*):

$$C = \{\mu \in R^T : \mu_1 \geq \mu_2 \geq \cdots \geq \mu_i \leq \mu_{i+1} \leq \mu_{i+2} \leq \cdots \leq \mu_n\}$$

(with at least one strict inequality among $\mu_1 \geq \mu_2 \geq \cdots \geq \mu$ and one among $\mu_{i+1} \leq \mu_{i+2} \leq \cdots \leq \mu_n$).

One may similarly define a cyclical profile with minima at $1, j$, and T and maxima at i and k. Also, we may similarly define an incomplete inequality profile.

Their method has several desirable properties, which make it a very strong method among its class. The estimator for the mean expression levels has several optimal properties, which are the same as that of Hwang and Peddada (1994). For example, the estimator universally dominates the unrestricted maximum likelihood estimator. Another important property is that the genes are selected into clusters based in part on a statistical significance criterion. Thus, it is possible to control type I error rates in this method, while it is not possible through the unsupervised methods such as cluster analytic algorithms. Also, another feature is that the method proposed by Peddada et al. (2003) can select genes with subtle but reproducible expression changes over time and, therefore, find some genes that may not be identified by other approaches. In the ORIOGEN approach, one needs only to describe the shapes of profiles in terms of mathematical inequalities; whereas, with the correlation-based procedures, one must specify numerical values at each time point for each candidate profile. ORIOGEN is designed for genes with a constant variance through time. This procedure is extended for hetroscedastic variances by Simmons and Peddada (2007), called ORIOGEN-Hetro, which uses an iterative algorithm to estimate mean expression at various time points when mean expression is subject to predefined order restrictions. This methodology lowers the false positive rate and keeps the same power as that of ORIOGEN. Both methodologies have been applied beautifully to a breast cancer cell-line data.

## CHAPTER 7

Bayesian Statistics

CONTENTS

7.1 BAYESIAN PROCEDURES

Traditionally, classical probabilistic approaches have been used in complex biological systems, which have high dimensionality with a high degree of noise and variability. Moreover, most of the variables in complex biological systems are either unknown or immeasurable. To make an inference, we must infer or integrate these variables by some suitable probabilistic models. The Bayesian probabilistic approach uses prior information, based on past knowledge about the possible values of parameters under consideration to give posterior distribution that contains all the probabilistic information about the parameters. The Bayesian approach has been used in bioinformatics, particularly in microarray data analysis such as the Mixture Model approach for gene expression, Empirical Bayes method, and Hierarchical Bayes method.

We described the Bayes theorem in Chapter 3. We will now explain the Bayes theorem in a simple way. Let us consider two events A and B and assume that both events have occurred. We now have two situations: Event A happened before event B, or event B happened before event A. Let us consider the first situation in which event A happened and then event B happened. Let the probability of the happening of event A be $P(A)$. Since event B happened after event A, the probability of the happening of event B is conditional on the happening of event A, and therefore, the probability of the happening of event B is given by $P(B|A)$. Since both events A and B have happened, the probability of the happening of both A and B is given by

$$P(A) \cdot P(B|A).$$

Let us now consider the second situation. The probability of the happening of event B is $P(B)$, and since event A happened after event B, the probability of the happening of event A is conditional upon the happening of event B. Therefore, the probability of event A given that event B has already happened is given by $P(A|B)$. Thus, the probability that both event A and B have happened is given by

$$P(B) \cdot P(A|B).$$

Since the sequences of events that A happened before B or A happened after B have the same probability,

$$P(A) \cdot P(B|A) = P(B) \cdot P(A|B). \tag{7.1.1}$$

Thus, we can use (7.1.1) to find the probability of B when A has already happened or to find the probability of A when B has already happened:

$$P(A|B) = \frac{P(A) \cdot P(B|A)}{P(B)}, \tag{7.1.2}$$

$$P(B|A) = \frac{P(B) \cdot P(A|B)}{P(A)}. \tag{7.1.3}$$

Equations (7.1.2) and (7.1.3) represent the Bayes theorem. The probabilities $P(A)$ and $P(B)$ are known as *prior* information, which is available about events A and B, respectively. The probabilities $P(A|B)$ and $P(B|A)$ are known as *posterior* probabilities of event A and B, respectively.

If we use H for denoting the hypothesized data and use O for denoting the observed data, then from (7.1.2) or (7.1.3), we get

$$P(H|O) = \frac{P(H) \cdot P(O|H)}{P(O)}. \tag{7.1.4}$$

The probability $P(O)$ can be written as

$$P(O) = P(H)P(O|H) + P(\overline{H})P(O|\overline{H}).$$

Here $\overline{H}$ is the complementary event of event H.

Thus, (7.1.4) can be written as

$$P(H|O) = \frac{P(H) \cdot P(O|H)}{P(H)P(O|H) + P(\overline{H})P(O|\overline{H})}. \qquad (7.1.5)$$

■ Example 1

Following is the R-code in which use of a beta prior is illustrated:

```
> Example 1
> # using a beta prior
library(LearnBayes)
q2 = list(p = .9,x = .5)
q1 = list(p = .4,x = .3)
c = beta.select(q1,q2)
alpha1 = c[1]
alpha2 = c[2]
s1 = 10
f1 = 15
curve(dbeta(x,alpha1 + s1,alpha2 + f1), from = 0, to = 1,
    xlab = "probability",ylab = "Density",lty = 1,lwd = 4)
curve(dbeta(x,s1 + 1,f1 + 1),add = TRUE,lty = 2,lwd = 4)
curve(dbeta(x,alpha1,alpha2),add = TRUE,lty = 3,lwd = 4)
legend(.9,4,c("Prior","Likelihood","Posterior"),
    lty = c(3,2,1),lwd = c(3,3,3))
posterior1 = rbeta(2000, alpha1 + s1, alpha2 + f1)
windows( )
hist(posterior1,xlab = "probability")
quantile(posterior1, c(0.05, 0.95))
```

■

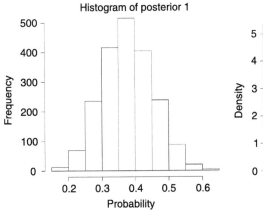
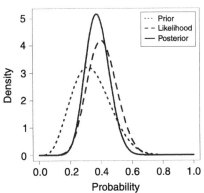

FIGURE 7.1.1

Prior, posterior, and likelihood using a beta distribution.

7.2 EMPIRICAL BAYES METHODS

Empirical Bayes methods allow drawing inferences on any single gene based on the information from other genes. Empirical Bayes methods were developed in Ecfron and Morris (1973); Efron et al. (2001); Robbins (1951); Robbins and Hannan (1955); and Efron (2006).

The principle of the empirical Bayes method is used for estimating the parameters of the mixture models of the following form:

$$f(v) = p_1 f_1(v) + p_0 f_0(v) \qquad (7.2.1)$$

where,

$p_1 = $ probability that a gene is affected,
$p_0 = 1 - p_1 = $ probability that a gene is unaffected,
$f_1(v) = $ the density of the gene expression V for affected genes,
$f_0(v) = $ the density of the gene expression V for unaffected genes.

To estimate $f(v)$, we need to estimate the density $f_0(v)$ for the unaffected gene expressions. In the absence of a strong parametric condition such as normality, it may not be easy to estimate $f_0(v)$.

Let us consider an experiment comparing the gene expression levels in two different situations such as condition 1 and condition 2. Let there be d_1 samples collected under condition 1 and d_2 samples collected under condition 2. Now we consider the difference between expression levels under condition 1 and condition 2. This may be done on the basis of some predefined conditions. Efron et al. (2001) used irradiated and unirradiated values within the same wild-type sample and aliquot to find the matrix D of difference between expression levels. From this matrix D, the null scores v_i are obtained. The scores v_i are used to estimate the null density $f_0(v)$.

The Bayes rule is used to find the posterior probability $p_0(v)$ and $p_1(v)$ that a gene with score V was in condition 1 or condition 2.

Thus, using the Bayes rule we get

$$p_1(V) = 1 - p_0 \frac{f_0(V)}{f(V)}, \qquad (7.2.2)$$

$$p_0(V) = p_0 \frac{f_0(V)}{f(V)}. \qquad (7.2.3)$$

The ratio $\frac{f_0(V)}{f(V)}$ is estimated directly from the empirical distributions. The bounds on the unknown probabilities p_0 and p_1 are obtained by using the following conditions:

$$p_1 \geq 1 - \min_v \left\{ \frac{f(V)}{f_0(V)} \right\}, \tag{7.2.4}$$

$$p_0 \leq \min_v \left\{ \frac{f(V)}{f_0(V)} \right\}. \tag{7.2.5}$$

The algorithm for empirical Bayes analysis is as follows:

a. Compute the scores $\{V_i\}$.

b. Compute the null scores $\{v_i\}$ using random permutation of labels in condition 1 and condition 2.

c. Use logistic regression to estimate the ratio $\frac{f_0(v)}{f(v)}$.

d. Use (7.2.4) and (7.2.5) to obtain the estimated upper bound for p_0.

e. Compute Prob $(Event|V)$, where *event* is condition 1 or condition 2, from (7.2.2) and (7.2.3) with $\frac{f_0(v)}{f(v)}$ estimated from the logistic regression, and p_0 equaling its estimated maximum value.

Since $\frac{f_0(v)}{f(v)}$ is estimated from the data itself instead of estimating it from an *a priori* assumption, it is called the empirical Bayes analysis. The empirical Bayes procedure provides an effective framework for studying the relative changes in the gene expression for a large number of genes. It employs a simple non-parametric mixture prior to modeling the population under condition 1 and condition 2, thereby avoiding parametric assumptions.

7.3 GIBBS SAMPLER

Recent developments in high-speed computing facilities have made computer-intensive algorithms powerful and important statistical tools. The Gibbs sampler was discussed by Geman and Geman (1984), who studied image-processing models. The actual work can be traced back to Metropolis et al. (1953). Later it was developed by Hastings (1970) and Gelfand and Smith (1990). Casella and George (1992) gave a simple and elegant explanation of how and why the Gibbs sampler works.

The principle behind the Gibbs sampler is to avoid difficult calculations, replacing them with a sequence of easier calculations. The Gibbs sampler is a technique for generating random variables from a distribution indirectly, without having to calculate the density. It is easy to see that Gibbs sampling is based on the elementary property of Markov chains.

Let us consider a joint density function

$$f(x, y_1, y_2, \ldots, y_p). \tag{7.3.1}$$

Suppose we are interested in finding the characteristics of the marginal density

$$g(x) = \int \cdots \int f(x, y_1, y_2, \ldots, y_p) dy_1, dy_2, \ldots, dy_p. \qquad (7.3.2)$$

The classical approach would be to find the marginal density (7.3.2) and then find the characteristics such as mean and variance. In most of the cases, the calculation of $g(x)$ may not be easy. In such cases, the Gibbs sampler is a natural choice for finding $g(x)$. The Gibbs sampler allows us to draw a random sample from $g(x)$ without requiring knowledge of the density function $g(x)$. The idea is taken from the fact that we can use a sample mean to estimate the population mean when sample size is large enough. In the same way, any characteristic of the population, even a density function, can be obtained using a large sample.

Let us consider a simple case of two variables (X, Y). Let the joint density of (X, Y) be denoted by $f(x, y)$ and the conditional distributions by $f(x|y)$ and $f(y|x)$. The Gibbs sampler generates a sequence of random variables called the "Gibbs sequence":

$$Y_0', X_0', Y_1', X_1', Y_2', X_2', \ldots, Y_k', X_k'. \qquad (7.3.3)$$

The initial value $Y_0' = y_0'$ is specified, and for obtaining the remaining values in (7.3.3), we use the following random-variable-generating scheme using conditional distributions:

$$X_j' \sim f\left(x|Y_j' = y_j'\right), \qquad (7.3.4)$$

$$Y_{j+1}' \sim f\left(y|X_j' = x_j'\right). \qquad (7.3.5)$$

Under reasonable conditions, the distribution of X_k' converges to $h(x)$ as $k \to \infty$. Thus, the sequence of random variables X_k' is in fact a random sample from the density function $h(x)$ for large k. This sequence of random variables is used to find characteristics of the distribution $h(x)$.

Suppose (X, Y) has the joint density

$$f(x, y) \propto \binom{n}{x} y^{x+\alpha-1} (1 - y)^{n-x+\beta-1}, x = 0, 1, \ldots, n; \ 0 \le y \le 1. \qquad (7.3.6)$$

To find the characteristics of the marginal distribution $h(x)$ of X, we generate a sample from $h(x)$ using the Gibbs sampler. For this, we notice that

$$f(x|y) \text{ is binomial } (n, y). \qquad (7.3.7)$$

$$f(y|x) \text{ is Beta } (x + \text{Beta}(x + \alpha, n - x + \beta). \qquad (7.3.8)$$

We use the Gibbs sampler scheme given in (7.3.4) and (7.3.5) to generate a random sample from $h(x)$ using (7.3.7) and (7.3.8). This sample is then used to estimate the characteristics of the distribution $h(x)$.

■ Example 1

Let us generate a bivariate normal sample using the Gibbs sampling scheme.

We can use R to generate a sample from a bivariate normal distribution with zero means, unit variances, and rho correlation between two variates as follows:

```
>> #Example 1
> #Gibbs sampling
> # bivariate normal
> gsampler<-function (n, rho1)
+ { mat <- matrix(ncol = 2, nrow = n)
+ x <- 0
+ y <- 0
+ mat[1, ] <- c(x, y)
+ for (i in 2:n) {
+ z<-sqrt(1-rho1^2)
+ x <- rnorm(1, rho1 * y, z)
+ y <- rnorm(1, rho1 * x, z)
+ mat[i, ] <- c(x, y)
+ }
+ mat
+ }
>
> bivariate1<-gsampler(20000,0.95)
> par(mfrow = c(3,2))
> plot(bivariate1,col = 1:20000)
> plot(bivariate1,type = "l")
> plot(ts(bivariate1[,1]))
> plot(ts(bivariate1[,2]))
> hist(bivariate1[,1],50)
> hist(bivariate1[,2],50)
> par(mfrow = c(1,1))
> # See Figure 7.3.1
```

■

The Gibbs sampling scheme has been extensively used in clustering of genes for extracting comprehensible information out of large-scale gene expression data sets. Clusters of coexpressed genes show specific functional categories and share cis-regulatory sequences in their promoters. In addition to approaches such

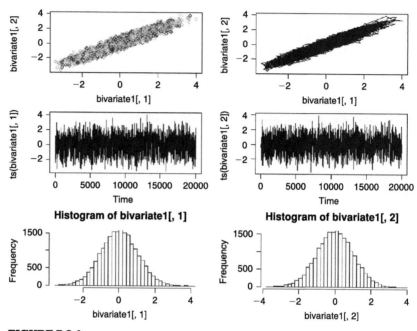

FIGURE 7.3.1

Gibbs sampling scheme.

as heuristic clustering methods, k-means, and self-organizing maps, there are some divisive and model-based clustering methods. A model-based clustering method assumes that the data set is generated by a mixture of distributions in which each cluster has its own distribution. This method takes into account the noise present in the data. The model-based clustering gives an estimate for the expected number of clusters and statistical inference about the clusters. The Gibbs sampling technique is used to make inference about the model parameters. The Gibbs sampling procedure is used iteratively to update the cluster assignment of each gene and condition using the full conditional distribution model. The following algorithm (Joshi et al., 2008) has been used to sample coclustering from the posterior distribution.

1. Randomly assign N genes to a random K_0 number of gene clusters, and for each cluster, randomly assign M conditions to a random $L_{k,0}$ number of condition clusters.

2. For N cycles, remove a random gene I from its current cluster. For each gene cluster k, calculate the Bayesian score $S(C_{i \to k})$, where $C_{i \to k}$ denotes the coclustering obtained from cocluster C by assigning gene i to cluster k, keeping all other assignments of genes and conditions equal, as well as the probability $S(C_{i \to 0})$ for the gene to be alone in its own

cluster. Assign gene i to one of the possible $K + 1$ gene clusters, where K is the current number of gene clusters, according to the probabilities $Q_k \propto e^{S(C_{i \to k})}$ normalized such that $\sum_k Q_k = 1$.

3. For each gene cluster k, for M cycles, remove a random condition m from its current cluster. For each condition cluster l, calculate the Bayesian score $S(C_{i \to k})$. Assign condition m to one of the possible $L_k + 1$ clusters, where L_k is the current number of condition clusters for gene cluster k, according to the probabilities $Q_1 \propto e^{S(C_{k,m \to 1})}$ normalized such that $\sum_1 Q_1 = 1$.

4. Repeat steps 2 and 3 until convergence. A Gibbs sampler is said to converge if two runs that start from different random initializations return the same averages for a suitable set of test functions.

For large-scale data sets, the posterior distribution is strongly peaked on a limited number of clusters with equal probability.

A transcriptional factor binding site binds transcription factors when a gene is expressed. RNA polymerase will not transcribe DNA if a transcription factor is binding. A specific transcription factor binds a specific binding site, and one can find the binding site by using phylogenetic footprinting, which allows the prediction of the binding site. Many genes participate in the same process at any given point of time; therefore, they co-express. A short motif may play an important role to bind transcription factors. Regulatory motifs usually have short fixed length and are repetitive even in a single gene, but vary extensively. One can use Gibbs sampling to find a motif.

Using the Gibbs sampling methods, researchers have increased the ability of procedures to detect differentially expressed genes and network connectivity consistency (Brynildsen et al., 2006). Even with the inaccuracy of ChiP-chip and gene expression data, it is possible to identify genes whose binding and gene expression data are closely related to each other more effectively by using the Gibbs sampler and the Bayesian statistics than those approaches that do not use Gibbs and Bayesian statistics. As compared to the model-based approach, one can look at the problem as a network problem. It is known that in bipartite network systems, various relationships exist between network components that result from the network topology. One can use the property that consistent genes are those genes that satisfy all network relationships between each other at a given level of tolerance. First, subnetworks need to be identified to act as seeds to find network relationships to identify consistent genes. Because consistent genes are not known in advance, the Gibbs sampler is used to populate these subnetworks with consistent genes, which will then maximize the number of consistent genes. The Gibbs sampler maximizes the function of all network components, which is considered to be a joint distribution. To maximize the joint distribution, a sample is iteratively chosen from a series of conditional distributions and select updates probabilistically. Seeds are selected randomly. The following algorithm (Brynildsen et al., 2006) is used:

1. Select a subnetwork β_i randomly for updating from a set of subnetworks.

2. Construct a conditional distribution $(\beta_i|\beta_{j\neq i}, Z_A, E)$ where Z_A is the connectivity pattern of matrix A, and E is the expression data, by examining all subnetworks β_i. This can be done by considering a set S_i of all the genes regulated by the transcription factors TF_i that are not current seeds for other transcription factors. Every gene that belongs to S_i is evaluated for its ability to identify consistent genes holding other subnetworks $\beta_{j\neq i}$ constant. This number of consistent genes identified by a different β_i is proportional to the probability that it is itself a consistent gene.

3. Normalize the conditional distribution and select updated β_i probabilistically.

Since the erred genes would be unlikely to be identified as consistent genes, this algorithm will select consistent genes to populate β_i. This method has an advantage over model-based procedures when >50% of the genes have erred connectivity, erred expression, or both.

For the identification of the differentially expressed genes between different conditions, several approaches have been proposed using the Bayes and empirical Bayes mixture analysis. Ishwaran and Rao (2003) used Bayesian ANOVA for microarray (BAM), which provides shrinkage estimates of gene effects derived from a form of generalized ridge regression. BAM uses a variable selection method by recasting the microarray data problem in terms of an ANOVA model and therefore as a linear regression model. BAM can use different model estimators, which can be made to analyze the data in various ways. It selects significance level α cutoff values for different model estimators, which is critical for any method in finding the differentially expressed genes. BAM balances between identifying a large number of genes and controlling the number of falsely discovered genes.

For group g, the kth expression is denoted by $Y_{j,k,g}$, $k = 1, 2, \ldots, n_{j,g}$, for gene $j = 1, 2, \ldots, p$. If we have only two groups, then $g = 1$ will denote the control group and $g = 2$ will denote the treatment group. If group sizes are fixed, then we can write $n_{j,1} = N_1$ and $n_{j,2} = N_2$. Let $\varepsilon_{j,k,g}$ be an iid normal $(0, \sigma_0^2)$ measurement error. Then, the ANOVA model is written as

$$Y_{j,k,g} = \theta_{j,0} + \mu_{j,0} I_{\{g=2\}} + \varepsilon_{j,k,g}, \tag{7.3.9}$$
$$j = 1, \ldots, p; \; k = 1, 2, \ldots, n_{j,g}; \; g = 1, 2.$$

where I is the indicator variable and has value 1 if the condition is met; otherwise, it is zero. If $\mu_{j,0} \neq 0$, then genes are expressing differentially.

The equation (7.3.9) can be extended in different cases such as analysis of covariance (ANCOVA) and hetroscedastic gene measurement. It can be shown that the assumption of normality in (7.3.9) is not necessary as long as the error measurement $\varepsilon_{j,k,g}$ is independent with finite moments and group sizes $n_{j,g}$ are large so that the central limit theorem can be applied. Now, the $Y_{j,k,g}$ can be replaced by its centered value

$$Y_{j,k,g} - \overline{Y}_{j,g} \tag{7.3.10}$$

where $\overline{Y}_{j,g}$ is the mean expression level of gene j over group g. This will force $\theta_{j,0}$ to be near zero, which will reduce the dimension to be p and correlation between the Bayesian parameters θ_j and μ_j.

Now $Y_{j,k,g}$ can be rescaled by using the unbiased σ_n^2 estimate of σ_0^2. Here

$$\hat{\sigma}_n^2 = \frac{1}{n - 2p} \sum_{j,k,g} \left(Y_{j,k,g} - \overline{Y}_{j,1} I_{\{g=1\}} - \overline{Y}_{j,2} I_{\{g=2\}}\right)^2. \tag{7.3.11}$$

Thus, the re-scaled $Y_{j,k,g}$ is

$$\widetilde{Y}_{j,k,g} = \left(Y_{j,k,g} - \overline{Y}_{j,g}\right) / \sqrt{n/\hat{\sigma}_n^2}. \tag{7.3.12}$$

where $n = \sum_{j=1}^{p} n_j$ is the total number of observations and $n_j = n_{j,1} + n_{j,2}$ is the total sample size for gene j.

After preprocessing, we can write (7.3.9) as

$$\widetilde{Y} = \widetilde{X}^T \widetilde{\beta}_0 + \widetilde{\varepsilon} \tag{7.3.13}$$

where $\widetilde{Y}$ is the vector of the n strung-out $\widetilde{Y}_{j,k,g}$ values, $\widetilde{\beta}_0$ is the new regression coefficient under the scaling, $\widetilde{\varepsilon}$ is the new vector of measurement errors, and $\widetilde{X}$ is the $n \times (2p)$ design matrix. The string-out is done by starting with the observations for the gene $j = 1$, group $g = 1$, followed by values of gene $j = 1$, group $g = 2$, followed by the values of gene $j = 2$, group $g = 1$, and so on. For β_j, we use the coefficients (θ_j, μ_j). Hierarchical prior variances for (θ_j, μ_j) are denoted by (v_{2j-1}^2, v_{2j}^2).

Ishwaran and Rao (2003) found the conditional distribution for β. For identifying parameters $\mu_{j,0}$ that are not zero, we use a Z-cut procedure. A gene j is identified as a differentially expressed gene if

$$|E(\mu_j|Y)| \geq z_{\frac{\alpha}{2}} \sqrt{\frac{n_j}{n_{j,1}}}$$

where $z_{\frac{\alpha}{2}}$ is the $100x\,(1-\alpha/2)$th percentile of a standard normal distribution. The value of $E(\mu_j|Y)$ is obtained by averaging Gibbs sampled values of μ_j.

Ishwaran and Rao (2005a) suggested a Bayesian rescaled spike and slab hierarchical model designed specifically for multigroup gene detection problems. The data preprocessing is done in a way so that it can increase the selection performance. The multigroup microarray problem is transformed in terms of an ANOVA framework, and then it uses a high-dimensional orthogonal model after a simple dimension-reduction and rescaling step. The transformed data are then modeled using a Bayesian spike and slab method. The ANOVA approach allows the estimation of all gene-group differential effects simultaneously and does not need pairwise group comparison to detect the differentially expressed genes across the experimental groups. With the use of rescaled spike and slab approach, the false detection rate is reduced.

Let the gene expression values for gene j from the ith microarray chip be denoted by $Y_{i,j} \cdot i = 1, \ldots, n$ and $j = 1, \ldots, M$. Each chip is assigned a group label $G_i \in \{1, \ldots, g\}$. Let g be the baseline group. Let $I(\cdot)$ denote the indicator function and let $n_k = \#(i : G_i = k)$ denote the number of samples from group k. Also, $n = \sum_{i=1}^{k} n_k$ is the total sample size for each gene. The multigroup ANOVA is then defined as

$$Y_{i,j} = \theta_j + \sum_{k=1}^{g-1} \beta_{k,j}\, I[G_i = k] + \varepsilon_{i,j}. \qquad (7.3.14)$$

Here θ_j, $\sum_{k=1}^{g-1} \beta_{k,j} I[G_i = k]$ and $\varepsilon_{i,j}$ are the baseline effect of gene j, gene-group interaction, and independent error, respectively. The errors are assumed to be independent with $E(\varepsilon_{i,j}) = 0$ and $E(\varepsilon_{i,j}^2) = \sigma_j^2$. The $\beta_{k,j}$'s measure the difference in mean expression level relative to baseline. A nonzero value of $\beta_{k,j}$ indicates a relative change in the mean expression value. The ANOVA model is converted to a rescaled spike and slab model after using a weighted regression technique to transform the data so that the data have equal variance and after reducing the dimension of the model. The rescaled spike and slab multigroup model is given by

$$\left(\mathbf{Y}_j^* | \boldsymbol{\beta}_j, \sigma^2\right) \sim N\left(\mathbf{X}_j\boldsymbol{\beta}_j, N\sigma^2\mathbf{I}\right), \quad j = 1, \ldots, M,$$

$$(\boldsymbol{\beta}_j | \boldsymbol{\gamma}) \sim N(\mathbf{0}, \boldsymbol{\Gamma}_j),$$

$$\boldsymbol{\gamma} \sim \pi(d\boldsymbol{\gamma}),$$

$$\sigma^2 \sim \mu(d\sigma^2). \qquad (7.3.15)$$

where $\boldsymbol{\Gamma}_j$ is the diagonal matrix with diagonal entries obtained from $\boldsymbol{\gamma}_j = (\gamma_{1,j}, \ldots, \gamma_{g-1,j})^t$. The design matrix X_j for gene j is chosen to satisfy orthogonality.

In a Bayesian probabilistic framework, Baldi and Long (2001) log-expression values are modeled by independent normal distributions, parameterized by corresponding mean and variances with hierarchical prior distributions. In this situation, all genes are assumed independent from each other, and they follow a continuous distribution such as the Gaussian distribution. Also, it is assumed that the expression values follow the Gaussian distribution $N(x; \mu, \sigma^2)$. For each gene, we have two parameters (μ, σ^2). The likelihood of the data D is given by

$$P(D|\mu, \sigma^2) = \prod_{i=1}^{n} N\left(x_i; \mu, \sigma^2\right)$$

$$= C(\sigma^2)^{-n/2} e^{-\sum_{i=1}^{n} (x_i - \mu)^2 / 2\sigma^2}. \qquad (7.3.16)$$

For microarray data, the conjugate prior is considered to be a suitable choice because it is flexible and suitable when both (μ, σ^2) are not independent. It may be noted that when both prior and posterior priors have the same distributional form, the prior is said to be a conjugate prior. For the gene expression data, Dirichlet distribution (Baldi and Brunak, 2001) is considered to be a good conjugate prior, which leads to the prior density of the form $P(\mu|\sigma^2)P(\sigma^2)$, where the marginal $P(\sigma^2)$ is a scaled inverse gamma and conditional $P(\mu|\sigma^2)$ is normal. This leads to a hierarchical model with a vector of four hyperparameters for the prior $\alpha = (\mu, \lambda_0, \nu_0, \sigma_0^2)$ with the densities

$$P(\mu|\sigma^2) = N(\mu; \mu_0, \sigma^2/\lambda_0), \qquad (7.3.17)$$

and

$$P(\sigma^2) = I\left(\sigma^2; \nu_0, \sigma_0^2\right). \qquad (7.3.18)$$

The prior has finite expectation if and only if $\nu_0 > 2$. The prior $P(\mu, \sigma^2) = P(\mu, \sigma^2|\alpha)$ is written as

$$C\sigma^{-1}(\sigma^2)^{-(\nu_0/2+1)} \exp\left[-\frac{\nu_0}{2\sigma^2}\sigma_0^2 - \frac{\lambda_0}{2\sigma^2}(\mu_0 - \mu)^2\right].$$

Here C is a normalizing coefficient. The hyperparameter μ_0 and σ^2/λ_0 can be viewed as the location and scale of μ. The hyperparameter ν_0 and σ_0^2 can be viewed as degrees of freedom and scale of σ^2. Using the Bayes theorem, we obtain the posterior distribution as follows:

$$P(\mu, \sigma^2|D, \alpha) = N\left(\mu; \mu_n, \sigma^2/\lambda_n\right) I\left(\sigma^2; \nu_n, \sigma_n^2\right), \qquad (7.3.19)$$

with

$$\mu_n = \frac{\lambda_0}{\lambda_0 + n}\mu_0 + \frac{n}{\lambda_0 + n}m,$$

$$\lambda_n = \lambda_0 + n,$$

$$\nu_n = \nu_0,$$

$$\nu_n \sigma_n^2 = \nu_0 \sigma_0^2 + (n - 1)s^2 + \frac{\lambda_0 n}{\lambda_0 + n}(m - \mu_0)^2.$$

Here m is the sufficient statistic for μ.

It can be shown that the conditional posterior distribution of mean $P(\mu | \sigma^2, D, \alpha)$ is normal $N(\mu_n, \sigma^2/\lambda_n)$, the marginal posterior of mean $P(\mu | D, \alpha)$ is Student's t, and the marginal posterior $P(\sigma^2 | D, \alpha)$ of the variance is scaled inverse gamma $I(\nu_n, \sigma_n^2)$.

The posterior distribution contains information about the parameters (μ, σ^2). The posterior estimates of the parameters (μ, σ^2) can be obtained by finding the mean of the posterior distribution or by finding the mode of the posterior distribution. These estimates are called the *mean of the posterior* (MP) or *maximum a posteriori* (MAP). In practice, both estimates give similar results when used on the gene expression data.

The MP estimate is given by

$$\mu = \mu_n \text{ and } \sigma^2 = \frac{\nu_n}{\nu_n - 2}\sigma_n^2 \text{ provided } \nu_n > 2.$$

If we take $\mu_0 = m$, then we get

$$\mu = m,$$

and

$$\sigma^2 = \frac{\nu_n}{\nu_n - 2}\sigma_n^2 = \frac{\nu_0 \sigma_0^2 + (n - 1)s^2}{\nu_0 + n - 2}, \tag{7.3.20}$$

provided $\nu_0 + n > 2$.

Here s^2 is the sufficient statistic for σ^2.

Baldi and Long (2001) suggested using the t-test with the regularized standard deviation given by (7.3.20). The number of corresponding degrees of freedom is associated with corresponding augmented population points.

Markov Chain Monte Carlo

CONTENTS

Markov chains have been widely used in bioinformatics in various ways. Markov Chain Monte Carlo (MCMC) is used mostly for optimization and overcoming problems faced when traditional methods are used. MCMC methods allow inference from complex posterior distributions when it is difficult to apply analytical or numerical techniques. Husmeier and McGuire (2002) used the MCMC method to locate the recombinant breakpoints in DNA sequence alignment of a small number of texa (4 or 5). Husmeier and McGuire (2002) used the MCMC method to make an inference about the parameters of the fitted model of the sequence of phylogenetic tree topologies along a multiple sequence alignment. Arvestad et al. (2003) developed a Bayesian analysis based on the MCMC which facilitated approximation of an *a posteriori* distribution for reconciliations. Thus, one can find the most probable reconciliations and estimate the probability of any reconciliation, given the observed gene tree. This method also gives an estimate of the probability that a pair of genes is ortholog. Orthology provides the most fundamental correspondence between genes in different genomes. Miklós (2003) used the MCMC method to genome rearrangement based on a stochastic model of evolution, which can estimate the number of different evolutionary events needed to sort a signed permutation.

To understand the MCMC methods, we need to understand the theory of the Markov chain, which governs the behavior of the MCMC methods.

8.1 THE MARKOV CHAIN

Let us consider a box divided into four intercommunicating compartments, labeled 1, 2, 3, and 4. These four compartments are the possible states of a system at any time, and set $S = \{1, 2, 3, 4\}$ is called the state space. Suppose a rabbit is placed in the box and let the random variable X denote the presence of the rabbit in the compartment. A movement from compartment i to j is called a transition of the system from state i to state j. If $i = j$, then we say there is no movement. The transition from compartment i to compartment j depends on the fact that the rabbit has to be in compartment i before it can go to compartment j, but the transition to compartment j does not depend on where the rabbit was before reaching compartment i. Thus, the probability that the rabbit moves to the next compartment depends only on where the rabbit was immediately before the move, not on which compartments the rabbit has visited in the past. This sequence of random variables defines the Markov chain.

Let's consider another situation. A simple unbiased coin is tossed repeatedly for a number of times with two possible outcomes, heads or tails, in each trial. Let p be the probability of getting heads and q be the probability of getting tails, $q + p = 1$. Let us denote heads by 1 and tails by 0 so that we can have a sequence of ones and zeros. Let X_n denote the outcome obtained at the nth toss of the coin. Then for $n = 1, 2, 3, \ldots$

$$P[X_n = 1] = p, \ P[X_n = 0] = q. \tag{8.1.1}$$

Since trials are independent, the outcome of the nth trial does not depend on outcomes of any of the previous trials. Thus, the sequence of random variables is independent and defies the Markov chain.

Now let's consider a partial sum of a random variable, $S_n = X_1 + X_2 + \cdots + X_n$. Thus, S_n gives the number of heads obtained in the first n trials with possible values $0, 1, 2, \ldots, n$.

We can see the relationship that

$$S_{n+1} = S_n + X_{n+1}. \tag{8.1.2}$$

If $S_n = j(j = 0, 1, 2, \ldots, n)$, then the random variable S_{n+1} can have only one of the two possible values. S_{n+1} can have the value j with the probability $j + 1$ with the probability p, and S_{n+1} can have the value j with the probability q. The probability that S_{n+1} assumes the value j or $j + 1$ depends on the variable S_n, not on $S_{n-1}, S_{n-2}, \ldots, S_1$.

Thus, we have

$$P[S_{n+1} = j + 1 | S_n = j] = p,$$
$$P[S_{n+1} = j | S_n = j] = q. \tag{8.1.3}$$

Therefore, we have a random variable S_{n+1}, which has a probability dependent only on the nth trial. The conditional probability states that the probability of S_{n+1} given S_n depends only on the S_n, not on the way the value of S_n has been achieved.

The Markov chain is defined as follows.

Definition *The stochastic process* $\{X_n, n = 0, 1, 2, \ldots\}$ *is called a Markov chain if, for* $j, k, i_1, i_2, \ldots, i_{n-1} \in N$ *(or any subset of integers I),*

$$P\left[X_n = k|X_{n-1} = j, X_{n-2} = i_1, X_{n-3} = i_2, \ldots, X_0 = i_{n-1}\right]$$
$$= P\left[X_n = k|X_{n-1} = j\right] = p(j, k). \qquad (8.1.4)$$

The outcomes of the Markov chain are called the states of the Markov chain. For example, if X_n has the outcome j, i.e., $X_n = j$, then the Markov chain is said to be at state j at the nth trial. The transition probabilities $p(j, k)$ describe the evolution of the Markov chain. The element $p(j, k)$ gives the probability that the chain at time n in state k, given that at time $n - 1$, it was in state j. This represents a conditional probability where k is stochastic and j is fixed. The transition probabilities can be arranged in a matrix form $\mathbf{P} = p(j, k)$ where i is a row suffix and j is a column suffix. We assume that the transition matrix is independent of time n, which means that the probability of a transition from state j to state k depends on two states, j and k, and not on time n. The transition matrix $\mathbf{P}$ is a stochastic matrix, which means that every entry of $\mathbf{P}$ satisfies $p(j, k) \geq 0$ and every row of $\mathbf{P}$ satisfies $\sum_k p(j, k) = 1$ (row sum is 1).

Consider a chain with state space $S = \{0, 1, 2\}$. The matrix $\mathbf{P}$ of order 3×3 takes the form

$$\mathbf{P} = \begin{bmatrix} p(0, 0) & p(0, 1) & p(0, 2) \\ p(1, 0) & p(1, 1) & p(1, 2) \\ p(2, 0) & p(2, 1) & p(2, 2) \end{bmatrix}$$

$$\mathbf{P} = \begin{bmatrix} \dfrac{1}{2} & 0 & \dfrac{1}{2} \\ \dfrac{1}{3} & \dfrac{1}{3} & \dfrac{1}{3} \\ \dfrac{1}{2} & \dfrac{1}{2} & 0 \end{bmatrix}.$$

In row 2, $P(1, 0) = 1/3, P(1, 1) = 1/3$, and $P(1, 2) = 1/3$, which gives the transition probability from state $\mathbf{X}_n = 1$ to state $\mathbf{X}_{n+1} = k$.

The probability of moving from state j to k in n transitions is given by

$$p^{(n)}(j, k) = P\left(X_n = k|X_0 = j\right) = P\left(X_{n+k} = k|X_n = j\right). \qquad (8.1.5)$$

The element $p^{(n)}(j, k)$ is the $(j + 1, k + 1)$ entry of $\mathbf{P}^n$.

Thus,

$$p^{(2)}(1, 2) = P(X_{2+n} = 2 | X_n = 1) = \sum_{k=0}^{2} P(X_{2+n} = 2, X_{1+n} = k | X_n = 1)$$

$$= \sum_{k=0}^{2} P(X_{1+n} = k | X_n = 1) P(X_{2+n} = 2 | X_{1+n} = k)$$

$$= p(1, 0)p(0, 2) + p(1, 1)p(1, 2) + p(1, 2)p(2, 2).$$

$$P(X_2 = 1 | X_0 = 2) = \frac{1}{2}.$$

$$P(X_2 = 2 | X_1 = 1) = \frac{1}{3}.$$

$$P(X_2 = 2, X_1 = 1 | X_0 = 2)$$

$$= P(X_2 = 2 | X_1 = 1). P(X_1 = 1 | X_0 = 2)$$

$$= \frac{1}{3} \cdot \frac{1}{2} = \frac{1}{6}.$$

The generalized result is given by

$$\mathbf{P}^{m+n} = \mathbf{P}^m \mathbf{P}^n. \tag{8.1.6}$$

In the use of Markov chains in sequence analysis, the concept of "time" is replaced with "position along the sequence." Our interest is to see whether the probability that a given nucleotide is present at any site depends on the nucleotide at the preceding site. Suppose $p(j, k)$ is the proportion of times that a nucleotide of type j is followed by a nucleotide of type k in a given DNA sequence. Then the data can be shown as follows:

nucleotide at site $i + 1$

		a	g	c	t
	a	$p(0, 0)$	$p(0, 1)$	$p(0, 2)$	$p(0, 3)$
nucleotide at site i	g	$p(1, 0)$	$p(1, 1)$	$p(1, 2)$	$p(1, 3)$
	c	$p(2, 0)$	$p(2, 1)$	$p(2, 2)$	$p(2, 3)$
	t	$p(3, 0)$	$p(3, 1)$	$p(3, 2)$	$p(3, 3)$

Graphically, a Markov chain is a collections of "states," each of which corresponds to a particular residue, with arrows between states. A Markov chain for DNA sequence is shown in Figure 8.1.1.

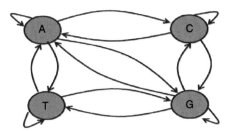

FIGURE 8.1.1
Markov chain for DNA sequence.

Each letter represents a state. The transition probability, shown by the arrow, is associated with each edge, which determines the probability of a certain letter following another letter or one state following another state.

The probability $p_{jk} = P(x_i = k | x_{i-1} = j)$ denotes the transition probability from state j to k in a single step.

For any probabilistic model of sequence, the probability of a DNA sequence x of length L is written as

$$P(x) = P(x_L, x_{L-1}, \ldots, x_1)$$
$$= P(x_L | x_{L-1}, \ldots, x_1) P(x_{L-1} | x_{L-2}, \ldots, x_1) \ldots P(x_1)$$
$$= P(x_L | x_{L-1}) P(x_{L-1} | x_{L-2}) \ldots P(x_2 | x_1) P(x_1)$$
$$= P(x_1) \prod_{i=2}^{L} P(x_i | x_{i-1}). \qquad (8.1.7)$$

It can be seen that the nucleotides at adjoining DNA sites are often dependent. The first-order Markov model fits better than the independence model on the real data.

■ Example 1

Consider a first-order Markov chain with the initial probability distribution $\{p_a = 0.1, p_c = 0.2, p_g = 0.4, p_t = 0.3\}$ and the transition matrix

nucleotide at site $i + 1$

		a	g	c	t
	a	0.1	0.2	0.1	0.6
nucleotide at site i	g	0.2	0.1	0.3	0.4
	c	0.3	0.3	0.2	0.2
	t	0.1	0.2	0.3	0.4

Find the probability of finding a DNA sequence CGGAT.

Solution

Using (8.1.7), we get

$$P(x) = P(x_5 = T, x_4 = A, x_3 = G, x_2 = G, x_1 = C)$$
$$= P(x_5|x_4)P(x_4|x_3)P(x_3|x_2)P(x_2|x_1)P(x_1 = C)$$
$$= p(A, T)p(G, A)p(G, G)p(C, G)p(C)$$
$$= 0.6 \cdot 0.2 \cdot 0.1 \cdot 0.3 \cdot 0.2$$
$$= 0.00072.$$

A major application of the equation (8.1.7) is in the calculation of the values of the likelihood ratio test.

The R-code for example 1 is as follows:

```
> #Example 1
> rm(list = ls( ))
> ttable<-matrix(c(0.1, .2, .1, .6, .2, .1, .3, .4, .3, .3, .2,
+ .2, .1, .2, .3, .4),nrow = 4,byrow = T)
> print("First order Markov Chain Probability matrix")
[1] "First order Markov Chain Probability matrix"
> rownames(ttable)<- c("a", "g", "c","t")
> colnames(ttable)<- c("a", "g", "c", "t")
> print("Transition Matrix")
[1] "Transition Matrix"
> ttable
  a   g   c   t
a 0.1 0.2 0.1 0.6
g 0.2 0.1 0.3 0.4
c 0.3 0.3 0.2 0.2
t 0.1 0.2 0.3 0.4
> pc<-0.2
> px<-ttable[1,4]*ttable[2,1]*ttable[2,2]*ttable[3,1]*pc
> cat("Probability of finding a DNA sequence CGGAT =", px, fill = T)
Probability of finding a DNA sequence CGGAT = 0.00072
```

■

■ Example 2

In the human genome, a C nucleotide is chemically modified by methylation wherever dinucleotide CG (generally written as CpG to distinguish it from the base-pair C-G) occurs. Due to this methylation, there is a high probability that methyl-C will mutate into T. As a consequence of this, it is rare to find dinucleotide CpG in the genome than would be expected from the independent probabilities of C and G. The methylation process is suppressed at some specific locations, such as regions around "promoters" and at the start of the sequence. These regions contain more dinucleotide CpG than anywhere else. These regions are called CpG islands. Suppose that from a set of human genomes, we find 45CpG islands. We consider two types of regions: One region is labeled as the region containing the CpG islands and the other region does not contain the CpG island. The transition probabilities are calculated by using the equation

$$p_{jk} = \frac{C_{jk}}{\sum_k C_{jk}} \qquad (8.1.8)$$

where C_{jk} is the number of times letter j followed letter k in the labeled regions. The p_{jk} gives the maximum likelihood estimates of the transition probabilities.

Let the transition probabilities based on approximately 65,000 nucleotides, calculated using (8.1.8), for model 1 (regions containing CpG islands) and model 2 (regions not containing CpG islands), be given as follows:

nucleotide at site $i + 1$

Model − 1	a	g	c	t
a	0.174	0.280	0.415	0.131
g	0.169	0.355	0.269	0.207
c	0.162	0.325	0.372	0.141
t	0.081	0.345	0.373	0.201

(nucleotide at site i labels the rows)

nucleotide at site $i + 1$

Model − 2	a	g	c	t
a	0.301	0.210	0.275	0.214
g	0.321	0.295	0.076	0.308
c	0.242	0.243	0.302	0.213
t	0.175	0.245	0.295	0.285

(nucleotide at site i labels the rows)

Find the log-odd ratio for the island CpG.

Solution

In both models, the probability that nucleotide G follows nucleotide C is lower than that of nucleotide C following G. The effect is seen to be very strong in model 2.

The log-odd ratio for the sequence x is given as follows:

$$SEQ(x) = \frac{P(x|model-1)}{P(x|model-2)} = \frac{\log_2\left(p_{jk}^{model\ 1}\right)}{\log_2\left(p_{jk}^{model\ 2}\right)}.$$

$$= \sum_{i=1}^{L} \omega_{i,i+1}$$

Then,

nucleotide at site $i+1$

	ω	a	g	c	t
nucleotide at site i a		-0.791	0.415	0.594	-0.708
g		-0.926	0.267	1.824	-0.573
c		-0.579	0.419	0.301	-0.595
t		-1.111	0.494	0.338	-0.504

The log-odd ratio of the island CpG is 1.824, which is very high.

The R-code for example 2 is as follows:

```
> #Example 2
> rm(list = ls( ))
> ttable1<-matrix(c(0.174, .28, .415, .131, .169, .355, .269, .207,
+ .162, .325, .372, .141,
+ .081, .345, .373, .201),nrow = 4,byrow = T)
> print("First order Markov Chain Probability matrix-Model-1")
[1] "First order Markov Chain Probability matrix-Model-1"
> rownames(ttable1)<-c("a", "g", "c", "t")
> colnames(ttable1)<-c("a", "g", "c", "t")
> print("Transition Matrix-Model 1")
[1] "Transition Matrix-Model 1"
> ttable1
  a g c t
a 0.174 0.280 0.415 0.131
g 0.169 0.355 0.269 0.207
c 0.162 0.325 0.372 0.141
t 0.081 0.345 0.373 0.201
>
> ttable2<-matrix(c(0.301, .210, .275, .214,
+ .321, .295, .076, .308,
+ .242, .243, .302, .213,
+ .175, .245, .295, .285),
```

```
+ nrow = 4,byrow = T)
> print("First order Markov Chain Probability matrix-Model-2")
[1] "First order Markov Chain Probability matrix-Model-2"
> rownames(ttable2)<-c("a", "g", "c", "t")
> colnames(ttable2)<-c("a", "g", "c", "t")
> print("Transition Matrix-Model 2")
[1] "Transition Matrix-Model 2"
> ttable2
  a     g     c     t
a 0.301 0.210 0.275 0.214
g 0.321 0.295 0.076 0.308
c 0.242 0.243 0.302 0.213
t 0.175 0.245 0.295 0.285
>
> t11<-ttable1[1,1]
> t21<-ttable2[1,1]
>
> r1c1<-(log2(t11/t21))
>
> t12<-ttable1[1,2]
> t22<-ttable2[1,2]
>
> r1c2<-(log2(t12/t22))
>
> t13<-ttable1[1,3]
> t23<-ttable2[1,3]
>
> r1c3<-(log2(t13/t23))
>
> t14<-ttable1[1,4]
> t24<-ttable2[1,4]
>
> r1c4<-(log2(t14/t24))
>
>
> t11<-ttable1[2,1]
> t21<-ttable2[2,1]
>
> r2c1<-(log2(t11/t21))
>
> t12<-ttable1[2,2]
> t22<-ttable2[2,2]
>
> r2c2<-(log2(t12/t22))
>
```

```
> t13<-ttable1[2,3]
> t23<-ttable2[2,3]
>
> r2c3<-(log2(t13/t23))
>
> t14<-ttable1[2,4]
> t24<-ttable2[2,4]
>
> r2c4<-(log2(t14/t24))
>
> t11<-ttable1[3,1]
> t21<-ttable2[3,1]
>
> r3c1<-(log2(t11/t21))
>
> t12<-ttable1[3,2]
> t22<-ttable2[3,2]
>
> r3c2<-(log2(t12/t22))
>
> t13<-ttable1[3,3]
> t23<-ttable2[3,3]
>
> r3c3<-(log2(t13/t23))
>
> t14<-ttable1[3,4]
> t24<-ttable2[3,4]
>
> r3c4<-(log2(t14/t24))
>
> t11<-ttable1[4,1]
> t21<-ttable2[4,1]
>
> r4c1<-(log2(t11/t21))
>
> t12<-ttable1[4,2]
> t22<-ttable2[4,2]
>
> r4c2<-(log2(t12/t22))
>
> t13<-ttable1[4,3]
> t23<-ttable2[4,3]
>
> r4c3<-(log2(t13/t23))
>
```

```
> t14<-ttable1[4,4]
> t24<-ttable2[4,4]
>
> r4c4<-(log2(t14/t24))
>
>
> ttable3<-matrix(c(r1c1, r1c2, r1c3, r1c4,
+ r2c1, r2c2, r2c3, r2c4,
+ r3c1, r3c2, r3c3, r3c4,
+ r4c1, r4c2, r4c3, r4c4),
+ nrow = 4, byrow = T)
> print("log-odd ratio Matrix")
[1] "log-odd ratio Matrix"
> rownames(ttable2)<-c("a", "g", "c", "t")
> colnames(ttable2)<-c("a", "g", "c", "t")
> print("Log-odd ratio matrix")
[1] "Log-odd ratio matrix"
> ttable3
[,1]    [,2]    [,3]    [,4]
[1,] -0.7906762 0.4150375 0.5936797 -0.7080440
[2,] -0.9255501 0.2671041 1.8235348 -0.5732996
[3,] -0.5790132 0.4194834 0.3007541 -0.5951583
[4,] -1.1113611 0.4938146 0.3384607 -0.5037664
> max<-max(ttable3)
>
> cat("Log-odd ratio of the island CpG is = ", max, fill = T)
Log-odd ratio of the island CpG is = 1.823535
>
```

8.2 APERIODICITY AND IRREDUCIBILITY

If the initial probability distribution and the transition probability matrix are given, then it is possible to find the distribution of the process at any specified time period n. The limiting distribution of the chain, independent of the initial starting distribution, converges to some invariant matrix. Suppose that $\pi'^{(n)}$ is the N-dimensional row vector denoting the probability distribution of X_n. The ith component of $\pi^{(n)}$ is

$$\pi^{(n)}(i) = P(X_n = i), \quad i \in S.$$

It can be shown that

$$\pi'^{(n)} = \pi'^{(n-1)} \mathbf{P}$$

and

$$\pi'^{(n)} = \pi'^{(0)}\mathbf{P}^n. \tag{8.2.1}$$

Then the limiting probability vector is given by

$$\pi' = \lim_{n\to\infty} \pi'^{(0)}\mathbf{P}^n.$$

The distribution π is said to be a stationary distribution if it satisfies

$$\pi' = \lim_{n\to\infty} \pi'^{(0)}\mathbf{P}^{n+1}$$

$$= \left(\lim_{n\to\infty} \pi'^{(0)}\mathbf{P}^n\right)\mathbf{P} \tag{8.2.2}$$

$$= \pi'\mathbf{P}.$$

Thus, when a chain reaches a point at which it has a stationary state, it retains its states for all subsequent moves. For finite state Markov chains, stationary distribution always exists. The period of state j is defined as the greatest common divisor of all integers $n \geq 1$ for which $p^{(n)}(j, j) > 0$. A chain is aperiodic if all states have a period 1. If all states of a Markov chain communicate such that every state is reachable from every other state in a finite number of transitions, then the Markov chain is called irreducible. If a finite Markov chain is aperiodic and irreducible with $n \geq 0$, and $\mathbf{P}^n$ has all entries positive, then that Markov chain is called an ergodic chain. An ergodic Markov chain converges to a unique stationary distribution π.

■ Example 1

Let us consider a three-state space Markov chain consisting of states $0, 1$, and 2 with the following transition matrix:

$$\mathbf{P} = \begin{bmatrix} p(0,0) & p(0,1) & p(0,2) \\ p(1,0) & p(1,1) & p(1,2) \\ p(2,0) & p(2,1) & p(2,2) \end{bmatrix}$$

$$= \begin{bmatrix} \frac{1}{4} & \frac{1}{4} & \frac{1}{2} \\ \frac{1}{4} & \frac{1}{2} & \frac{1}{4} \\ 0 & \frac{2}{3} & \frac{1}{3} \end{bmatrix}.$$

From this matrix, we can find the probability distribution of $\mathbf{X}_{n+1}$ given $\mathbf{X}_n$. Find the transition matrix after four transitions. If the starting probability of the chain is

$$\pi'^{(0)} = \begin{pmatrix} 0 & 0 & 1 \end{pmatrix}$$

find the stationary distribution.

Solution

The transition matrix after four transitions is obtained by taking the fourth power of the transition matrix; it is given by the following matrix:

$$\mathbf{P^4} = \begin{bmatrix} 0.172309 & 0.511429 & 0.316262 \\ 0.168837 & 0.512442 & 0.318721 \\ 0.171296 & 0.50733 & 0.321373 \end{bmatrix}.$$

The stationary distribution is obtained by using (8.2.2). For this, we multiply $\pi'^{(0)}$ with P to obtain $\pi'^{(1)}$.

$$\pi'^{(1)} = \begin{pmatrix} 0 & \dfrac{2}{3} & \dfrac{1}{3} \end{pmatrix}.$$

Similarly, we obtain $\pi'^{(2)}$ by multiplying $\pi'^{(1)}$ with P.

$$\pi'^{(2)} = \begin{pmatrix} 0.1702128 & 0.5106258 & 0.3191522 \end{pmatrix}.$$

In the same way,

$$\pi'^{(3)} = \begin{pmatrix} 0.170218 & 0.510663 & 0.319148 \end{pmatrix},$$

$$\pi'^{(4)} = \begin{pmatrix} 0.171297 & 0.5106383 & 0.3191489 \end{pmatrix}.$$

After the tenth iteration, we get the approximate stationary distribution as

$$\pi' = \begin{pmatrix} 0.170212 & 0.510639 & 0.319149 \end{pmatrix}. \qquad (8.2.3)$$

The same stationary distribution (8.2.3) can also be obtained regardless of values of the starting probability distribution $\pi'^{(0)}$.

We can also calculate the stationary distribution by using a system of linear equations (8.2.2) as follows.

Let the stationary distribution be denoted by

$$\pi' = \begin{pmatrix} \pi(0) & \pi(1) & \pi(2) \end{pmatrix},$$

with

$$\pi(2) = 1 - \pi(0) - \pi(1).$$

Now using (8.2.2), we can form the system of equations to solve for $\pi(0)$ and $\pi(1)$:

$$\frac{\pi(0)}{4} + \frac{\pi(1)}{4} + \frac{\pi(2)}{2} = \pi(0),$$

$$\frac{\pi(0)}{4} + \frac{\pi(1)}{2} + \frac{\pi(2)}{4} = \pi(1),$$

$$0 + \frac{2\pi(1)}{3} + \frac{\pi(2)}{3} = \pi(2).$$

We get

$$\pi(0) = \frac{1}{3}, \pi(1) = \frac{1}{3}, \pi(2) = \frac{1}{3}. \tag{8.2.4}$$

After a very large number of iterations, (8.2.3) will eventually converge to (8.2.4).

The following R-code calculates the approximate stationary distribution:

```
> #Example 1
> rm(list = ls( ))
> library(Biodem)
> p1<-matrix(c(1/4, 1/4, 1/2,
+ 1/4, 1/2, 1/4,
+ 0, 2/3, 1/3), nrow = 3, byrow = T)
> print("Transition Matrix-P")
[1] "Transition Matrix-P"
> p1
[,1]   [,2]   [,3]
[1,] 0.25 0.2500000 0.5000000
[2,] 0.25 0.5000000 0.2500000
[3,] 0.00 0.6666667 0.3333333
>
> pi01<-matrix(c(0, 0, 1),
+ nrow = 1,byrow = T)
> print("Starting Probability of the Chain")
[1] "Starting Probability of the Chain"
> pi01
[,1]   [,2]   [,3]
[1,]  0   0   1
>
> p4<-mtx.exp(p1, 4)
> print("4th power of Matrix P")
[1] "4th power of Matrix P"
> p4
[,1]   [,2]   [,3]
```

```
[1,] 0.1723090 0.5114294 0.3162616
[2,] 0.1688368 0.5124421 0.3187211
[3,] 0.1712963 0.5073302 0.3213735
>
> pi1<-pi01 %*% p4
> print("Probability PI(1)")
[1] "Probability PI'(1)"
> pi1
[,1]    [,2]    [,3]
[1,] 0.1712963 0.5073302 0.3213735
>
> pi2<-pi1 %*% p4
> print("Probability PI'(2)")
[1] "Probability  PI'(2)"
> pi2
[,1]    [,2]    [,3]
[1,] 0.170222 0.5106258 0.3191522
>
> pi3<-pi2 %*% p4
> print("Probability PI'(3)")
[1] "Probability PI'(3)"
> pi3
[,1]    [,2]    [,3]
[1,] 0.1702128 0.5106383 0.3191489
>
> pi4<-pi3 %* % p4
> print("Probability PI'(4)")
[1] "Probability PI'(4)"
> pi4
[,1]    [,2]    [,3]
[1,] 0.1702128 0.5106383 0.3191489
>
> pi5<-pi4 %*% p4
> print("Probability PI'(5)")
[1] "Probability PI'(5)"
> pi5
[,1]    [,2]    [,3]
[1,] 0.1702128 0.5106383 0.3191489
>
> pi6<-pi5 %*% p4
> print("Probability PI'(6)")
[1] "Probability PI'(6)"
> pi6
[,1]    [,2]    [,3]
[1,] 0.1702128 0.5106383 0.3191489
```

```
>
>
> pi7<-pi6 %*% p4
> print("Probability PI'(7)")
[1] "Probability PI'(7)"
> pi7
[,1]    [,2]    [,3]
[1,] 0.1702128 0.5106383 0.3191489
>
>
> pi8<-pi7 %*% p4
> print("Probability PI'(8)")
[1] "Probability PI'(8)"
> pi8
[,1]    [,2]    [,3]
[1,] 0.1702128 0.5106383 0.3191489
>
>
> pi9<-pi8 %*% p4
> print("Probability PI'(9)")
[1] "Probability PI'(9)"
> pi9
[,1]    [,2]    [,3]
[1,] 0.1702128 0.5106383 0.3191489
>
>
> pi10<-pi9 %*% p4
> print("Probability PI'(10)")
[1] "Probability PI'(10)"
> pi10
[,1]    [,2]    [,3]
[1,] 0.1702128 0.5106383 0.3191489
>
```

8.3 REVERSIBLE MARKOV CHAINS

Let an ergodic Markov chain with state space S converge to an invariant distribution π. If x and y denote the current and next state, and both $x, y \in S$, then a Markov chain is said to be reversible if it satisfies the condition

$$\pi(x)p(x, y) = \pi(y)p(y, x), \quad \text{for} \quad x, y \in S \qquad (8.3.1)$$

where $p(x, y)$ denotes the probability of a transition from state x to state y, and $p(y, x)$ denotes the probability of a transition from state y to x.

Equation (8.3.1) is called a balanced equation. It is of interest to note that an ergodic chain in equilibrium has its stationary distribution if it satisfies (8.3.1).

One of the most important uses of the reversibility condition is in the construction of MCMC methods. If a Markov chain needs to be generated with a stationary distribution, then one can use the reversibility condition as a point of departure.

■ Example 1

Let us consider the three-state chain on $S = \{0, 1, 2\}$ with the following transition probability matrix:

$$\mathbf{P} = \begin{bmatrix} 0.3 & 0.4 & 0.3 \\ 0.3 & 0.3 & 0.4 \\ 0.4 & 0.3 & 0.3 \end{bmatrix}.$$

For large n, we can find that

$$\mathbf{P}^n = \begin{bmatrix} \dfrac{1}{3} & \dfrac{1}{3} & \dfrac{1}{3} \\ \dfrac{1}{3} & \dfrac{1}{3} & \dfrac{1}{3} \\ \dfrac{1}{3} & \dfrac{1}{3} & \dfrac{1}{3} \end{bmatrix}.$$

Therefore, we can write the unique stationary distribution as

$$\pi = \begin{bmatrix} \dfrac{1}{3} & \dfrac{1}{3} & \dfrac{1}{3} \end{bmatrix}'.$$

Now we check the condition for reversibility (8.3.1):

$$\pi(0)p(0, 1) = \frac{1}{3}(0.4)$$

$$\neq \pi(1)p(1, 0) = \frac{1}{3}(0.3).$$

Therefore, this chain is not reversible.

Following is the R-code for example 1:

```
> #Example 1
> rm(list = ls( ))
> library(Biodem)
> p1<-matrix(c(0.3, 0.4, 0.3,
+ 0.3, 0.3, 0.4,
```

```
+ 0.4, 0.3, 0.3), nrow = 3, byrow = T)
> print("Transition Matrix-P")
[1] "Transition Matrix-P"
> p1
     [,1] [,2] [,3]
[1,] 0.3 0.4 0.3
[2,] 0.3 0.3 0.4
[3,] 0.4 0.3 0.3
>
> pn<-mtx.exp(p1, 100)
> print("nth power Transition matrix")
[1] "nth power Transition matrix"
> pn
          [,1]      [,2]      [,3]
[1,] 0.3333333 0.3333333 0.3333333
[2,] 0.3333333 0.3333333 0.3333333
[3,] 0.3333333 0.3333333 0.3333333
>
> piprime<-matrix(c(1/3, 1/3, 1/3), nrow = 1, byrow = T)
>
> r1<-piprime[1,1]*p1[1,2]
> r2<-piprime[1,2]*p1[2,1]
> if(r1! = r2) point ("Chain is not reversible")
[1] "Chain is not reversible"
>
```

■

8.4 MCMC METHODS IN BIOINFORMATICS

In many situations in bioinformatics, analytical or numerical integration techniques cannot be applied. Consider an m-dimensional random point $X = (X_1, X_2, \ldots, X_m)$. If $X_1, X_2, \ldots, X_m$ are independent, then one can generate random samples for each coordinate, but if $X_1, X_2, \ldots, X_m$ are dependent, then it becomes difficult to generate samples. When direct sampling from the distribution is computationally intractable, the Markov Chain Monte Carlo (MCMC) has become a very important tool to make inferences about complex posterior distributions. The main idea behind the MCMC methods is to generate a Markov chain using a Monte Carlo simulation technique. The generated Markov chain has desired posterior distribution as stationary distribution.

MCMC provides techniques for generating a sample sequence $X_1, X_2, \ldots, X_t$, where each $X_j = \{X_{1j}, X_{2j}, \ldots, X_{mj}\}$ has the desired distribution π. This is done by randomly generating X_j conditionally on X_{j-1}. With this technique, though the cost per unit to generate a sample is very low, it makes elements of

$X_1, X_2, \ldots, X_t$ correlated rather than independent. This results in large variances for sample averages and therefore results in less accuracy.

The classic papers of Metropolis et al. (1953) and Hastings (1970) gave a very general MCMC method under the generic name *Metropolis-Hastings algorithm*. The Gibbs sampler, discussed in Chapter 7, is a special case of Metropolis-Hastings algorithm. Metropolis et al. (1953) dealt with the calculation of properties of chemical substances. Consider a substance with m molecules positioned at $X = (X_1, X_2, \ldots, X_m)'$. Here, each X_j is formed by the two-dimensional vector positions in a plane of the jth molecule. The calculation of equilibrium value of any chemical property is achieved by finding the expected value of a chemical property with respect to the distribution of the vector of positions. If m is large, then the calculation of expectation is not possible. Metropolis et al. (1953) used an algorithm to generate a sample from the desired density function to find the Monte Carlo estimate of the expectation.

Hastings (1970) suggested an algorithm for the simulation of discrete equilibrium distributions on a space of fixed dimensions. If the dimension is not fixed, then one can use the reversible jump MCMC algorithm suggested by Green (1995), which is a generalization of the Metropolis-Hastings algorithm. The dimension of the model is inferred through its marginal posterior distribution.

MCMC applies to both continuous and discrete state spaces. We focus on the discrete version.

The Metropolis-Hastings algorithm first generates candidate draws from a distribution called a "proposal" distribution. These candidate draws are then "smoothed" so that they behave asymptotically as random observations from the desired equilibrium or target distribution. In each step of the Metropolis-Hastings algorithm, a Markov chain is constructed in two steps. These two steps are, namely, a proposal step and an acceptance step. These two steps are based on the proposal distribution (prior distribution) and on the acceptance probability.

At the nth stage, the state of the chain is $X_n = x$. Now to get the next state of the chain, sample a candidate point $Y_{n+1} = y$ from a proposal distribution with PDF $z(x, .)$. The PDF $z(x, .)$ may or may not depend on the present point x. If $z(x, .)$ depends on a present point x, then $z(x, y)$ will be a conditional PDF of y given x. The candidate point y is then accepted with some probability $c(x, y)$. The next state becomes $X_{n+1} = y$ if the candidate point y is accepted, and the next state becomes $X_{n+1} = x$ if the candidate point is rejected.

Now we need to find the probability $c(x, y)$ used in the Metropolis-Hastings algorithm. Following Chib and Greenberg (1995) and Waagepetersen and Sorensen (2001), we derive $c(x, y)$ as follows.

Let the random vector (X_n, Y_{n+1}) represent the current Markov chain state and proposal distribution point. Let the joint distribution of (X_n, Y_{n+1}) be $g(x, y)$ where

$$g(x, y) = z(x, y)\kappa(x), \tag{8.4.1}$$

and $\kappa(x)$ is the equilibrium PDF.

If z satisfies the reversibility condition such that

$$z(x, y)\kappa(x) = z(y, x)\kappa(y), \tag{8.4.2}$$

for every (x, y), then the proposal PDF $z(x, y)$ is the correct transition kemel (transition probability function) of the Metropolis-Hastings algorithm. Since the chain is reversible, we can now introduce the probability $c(x, y) < 1$, which is the probability that a candidate point y is accepted. Thus, (8.4.2) can be written as

$$z(x, y)\kappa(x)c(x, y) = z(y, x)\kappa(y)c(y, x). \tag{8.4.3}$$

Now set $c(y, x) = 1$; this will yield

$$z(x, y)\kappa(x)c(x, y) = z(y, x)\kappa(y). \tag{8.4.4}$$

From (8.4.3), we find the acceptance probability $c(x, y)$ as follows:

$$c(x, y) = \frac{z(y, x)\kappa(y)}{z(x, y)\kappa(x)}. \tag{8.4.5}$$

One can set $c(x, y) = 1$ in (8.4.3) to get the value of $c(y, x)$ in a similar way. Moreover, c guarantees that the equation is balanced. Thus, we have

$$c(x, y) = \min\left(1, \frac{z(y, x)\kappa(y)}{z(x, y)\kappa(x)}\right), \quad z(x, y)\kappa(x) > 0. \tag{8.4.6}$$

If the proposal densities are symmetric in nature, which means $z(y, x) = z(x, y)$, then (8.4.6) reduces to

$$c(x, y) = \min\left(1, \frac{\kappa(y)}{\kappa(x)}\right), \quad \kappa(x) > 0.$$

In the Metropolis-Hastings algorithm, the choice of parameterization and the proposal density play an important role. The choice of proper proposal densities is discussed by Chib and Greenberg (1995) and the optimal acceptance rate for high dimensional case is discussed by Roberts et al. (1977).

■ Example 1

Let us consider a single observation $\mathbf{y} = (y_1, y_2)$ from a bivariate normal distribution with the distribution function

$$f(\mathbf{y}|\boldsymbol{\mu}, C) = N(\boldsymbol{\mu}, C),$$

where

$$\boldsymbol{\mu} = (\mu_1, \mu_2), \quad \text{and} \quad C = \begin{bmatrix} 1 & \rho \\ \rho & 1 \end{bmatrix}.$$

Let the prior distribution, $h(\boldsymbol{\mu})$ of $\boldsymbol{\mu}$ be proportional to a constant, independent of $\boldsymbol{\mu}$ itself.

Using the Bayesian theory (Chapter 7), we find that the posterior distribution of $\boldsymbol{\mu}$ is given by

$$g(\boldsymbol{\mu}|\mathbf{y}, C) \propto h(\boldsymbol{\mu})f(\mathbf{y}|\boldsymbol{\mu}, C)$$

$$\text{or} \quad g(\boldsymbol{\mu}|\mathbf{y}, C) \propto f(\mathbf{y}|\boldsymbol{\mu}, C). \tag{8.4.7}$$

Thus, the posterior distribution is bivariate normal, which is the same as the distribution of observation $\mathbf{y}$.

According to the Gibbs sampling scheme (Chapter 7), one can draw observations using

$$w_1(\mu_1|\mu_2, \mathbf{y}, C) \sim N(y_1 + \rho(\mu_2 - y_2), 1 - \rho^2) \tag{8.4.8}$$

and

$$w_2(\mu_2|\mu_1, \mathbf{y}, C) \sim N(y_2 + \rho(\mu_2 - y_1), 1 - \rho^2). \tag{8.4.9}$$

We continue sampling using (8.4.8) and (8.4.9) until it converges to a stationary distribution. Thus, the samples (μ_1, μ_2) obtained using (8.4.8) and (8.4.9) are MCMC from $g(\boldsymbol{\mu}|\mathbf{y}, C)$.

The discovery of a high frequency of mosaic RNA sequences in HIV-1 suggests that a substantial proportion of AIDS patients have been coinfected with an HIV-1 strain belonging to different subtypes, and that recombination between these genomes can occur *in vivo* to generate new biologically active viruses (Robertson et al., 1995). It appears that the recombination, which is a horizontal transfer of DNA/RNA substances, may be an important source of genetic diversification in certain bacteria and viruses. The identification of recombination may lead to some new treatment and drug strategies. There are many phylogenetic methods available for detecting recombination. These methods are based on the fact that the recombination methods lead to a change of phylogenetic tree topology in the affected region of the sequence alignment. Thus, if the nonrecombinant part of the sequence alignment is known, it can be used as a reference to identify corresponding recombinant regions.

Husmeier and McGuire (2002) presented an MCMC method to detect the recombination whose objective is to accurately locate the recombinant breakpoints in DNA sequence alignment of small numbers of texa (4 or 5). Let there be an alignment D containing m DNA sequences with N nucleotides. Each column in the alignment is represented by y_t; t represents a site and $1 \leq t \leq N$. Thus, y_t is an m-dimensional column vector containing the nucleotides at the tth site of the alignment, and $D = (y_1, \ldots, y_N)$. Let $S = (S_1, \ldots, S_N)$ where S_t (the state at site t) represents the tree topology at site t. With each state S_t, there can be a vector of branch length, W_{S_t}, and nucleotide substitution parameters, θ_{S_t}. Let $\mathbf{w} = (w_1, \ldots, w_k)$ and $\theta = (\theta_1, \ldots, \theta_k)$.

Due to recombination, the tree topology is changed, which means that there is a transition from one state to another state S_t at the breakpoint t of the affected region. Thus, we would like to predict the state sequence $S = (S_1, \ldots, S_N)$. This prediction can be achieved by using the posterior probability $P(S|D)$, which is given by

$$P(\mathbf{S}|\mathbf{D}) = \int P(\mathbf{S}, \mathbf{w}, \boldsymbol{\theta}, v|D)\, dw\, d\theta\, dv. \qquad (8.4.10)$$

Here $v \in (0, 1)$ represents the difficulty of changing topology.

The joint distribution of DNA sequence alignment, the state sequence, and the model parameters are given by

$$P(\mathbf{D}, \mathbf{S}, \mathbf{w}, \boldsymbol{\theta}, v) = \prod_{t=1}^{N} P(y_t|S_t, \mathbf{w}, \boldsymbol{\theta}) x \prod_{t=1}^{N} P(S_t|S_{t-1}, v) P(S_1) P(\mathbf{w}) P(\boldsymbol{\theta}) P(v)$$

$$(8.4.11)$$

where $P(y_t|S_t, \mathbf{w}, \boldsymbol{\theta})$ is the probability of the tth column of the nucleotide in the alignment; $P(S_t|S_{t-1}, v)$ is the probability of transitions between states; and $P(\boldsymbol{\theta})$, $P(S_1)$, $P(\mathbf{w})$, and $P(v)$ are the prior probabilities.

Let the subscript i denote the ith sample from the Markov chain. Then using the Gibbs sampling procedure, sampling each parameter group separately conditionally on others, we obtain the $(i+1)$th sample as follows:

$$\mathbf{S}^{(i+1)} \sim P(.|\mathbf{w}^{(i)}, \boldsymbol{\theta}^{(i)}, v^{(i)}\mathbf{D}),$$
$$\mathbf{w}^{(i+1)} \sim P(.|\mathbf{S}^{(i+1)}, \boldsymbol{\theta}^{(i)}, v^{(i)}, \mathbf{D}),$$
$$\boldsymbol{\theta}^{(i+1)} \sim P(.|\mathbf{S}^{(i+1)}, \mathbf{w}^{(i+1)}, v^{(i)}, \mathbf{D}),$$
$$v^{(i+1)} \sim P(.|\mathbf{S}^{(i+1)}, \mathbf{w}^{(i+1)}, \boldsymbol{\theta}^{(i+1)}, \mathbf{D}). \tag{8.4.12}$$

Let K be the total number of states,

$$P(S_t|S_{t-1}, \ldots, S_1) = P(S_t|S_{t-1}) = v\delta(S_t, S_{t-1}) + \frac{1-v}{K-1}\left[1 - \delta(S_t, S_{t-1})\right], \tag{8.4.13}$$

where $\delta(S_t, S_{t-1})$ denotes the Kronecker delta function, which is 1 when $S_t = S_{t-1}$ and 0 otherwise.

Then,

$$P(v|\alpha, \beta) \propto v^{\alpha-1}(1-v)^{\beta-1}. \tag{8.4.14}$$

Define

$$\psi = \sum_{t=2}^{N} \delta(S_{t-1}, S_t).$$

Then, the joint probability function can be written as

$$P(\mathbf{D}, \mathbf{S}, \mathbf{w}, \boldsymbol{\theta}, v) \propto v^{\psi+\alpha-1}(1-v)^{N-\psi+\beta-2}.$$

On normalization, it gives a beta distribution

$$P(v|\mathbf{D}, \mathbf{S}, \mathbf{w}, \boldsymbol{\theta}) \propto v^{\psi+\alpha-1}(1-v)^{N-\psi+\beta-2}. \tag{8.4.15}$$

We can sample each state S_t separately conditional on the others as follows:

$$S_1^{(i+1)} \sim P(.|S_1^{(i)}, S_3^{(i)}, S_4^{(i)}, \ldots, S_N^{(i)}, \mathbf{w}^{(i)}, \boldsymbol{\theta}^{(i)}, v^{(i)}, \mathbf{D})$$

.
.
.

$$S_N^{(i+1)} \sim P(.|S_1^{(i+1)}, S_2^{(i+1)}, S_3^{(i+1)}, \ldots, S_{N-1}^{(i+1)}, \mathbf{w}^{(i)}, \boldsymbol{\theta}^{(i)}, v^{(i)}, \mathbf{D}).$$

For sampling the remaining parameters, $\mathbf{w}$ and $\boldsymbol{\theta}$, we can apply the Metropolis-Hastings algorithm. Let $\mathbf{Z}^{(i)}$ denote the parameter configuration in the ith sampling step; the new parameter configuration $\tilde{Z}$ is sampled from a proposal distribution $\mathbf{Q}(\tilde{Z}|\mathbf{Z}^{(i)})$, and the acceptance probability is given by

$$\mathbf{A}(\tilde{Z}) = \min \left\{ \frac{\mathbf{P}(\tilde{Z})\mathbf{Q}\left(\mathbf{Z}^{(i)}|\tilde{Z}\right)}{\mathbf{P}(\mathbf{Z}^{(i)})\mathbf{Q}(\tilde{Z}|\mathbf{Z}^{(i)})}, 1 \right\}, \qquad (8.4.16)$$

where P is given by (8.4.11).

Thus, Markov Chain Monte Carlo (MCMC) is used in bioinformatics for optimization and for overcoming some of the problems faced when traditional methods are used.

There are several packages available for MCMC procedures. For example, AMCMC (Rosenthal, 2007) is a software package for running adaptive MCMC algorithms on user-supplied density functions. It allows a user to specify a target density function and a desired real-value function. The package then estimates the expected value of that function with respect to that density function, using the adaptive Metropolis-within-Gibbs algorithm.

Analysis of Variance

Analysis of variance (ANOVA) is a powerful technique in the statistical analysis used in many scientific fields. In genomics, ANOVA is primarily used for testing the difference in several population means although there are several other uses of ANOVA. The t-test, discussed in Chapter 5, can be used only for testing the difference between two population means. The t-test is not applicable for testing the difference between three or more population means. Such situations are very common in genomics studies where we have more than two populations being studied. For example, a typical situation is whether there is any difference among expression levels of the given gene measured under k conditions. In such situations, we use the ANOVA technique. The objective of ANOVA is to test the homogeneity of several means. The ANOVA technique is formulated for testing the difference in several means, but the name of the technique seems to suggest that this technique is for the analysis of variance and not for the difference of means. The term *analysis of variance* was introduced by Professor R. A. Fisher in 1920 while working on a problem in the analysis of agronomical data.

The variations in the given data may be due to two types of causes—namely, assignable causes and chance causes. Assignable causes can be identified and can be removed, but chance causes are beyond control and cannot be removed. The idea behind ANOVA is to separate variance assignable to one group from variance assignable to another group. ANOVA provides information that allows us to draw the inference about the difference in means. If we have a certain difference in means, we can reject the null hypothesis that the population means are the same provided the variance from each sample is sufficiently small in comparison to the overall variances. On the other hand, if the variance from the samples is large as compared to the overall variance, then the null hypothesis that the population means are equal is not rejected.

Following are the assumptions needed for the F-test used in the ANOVA technique:

1. All observations are independent.

2. Parent populations from which the samples are drawn are normal.

3. Various effects are additive in nature.

Since the ANOVA technique assumes that the parent population from which samples are drawn is normal, we find that ANOVA is a parametric test.

The main objective of ANOVA is to examine whether there is a significant difference between class means in the presence of the inherent variability within the different classes.

9.1 ONE-WAY ANOVA

Let us consider that there are N observations $X_{ij}, i = 1, 2, \ldots, k; j = 1, 2, \ldots, n_i$ of a random variable X grouped (known as treatments or effects or conditions) into k classes of sizes $n_1, n_2, \ldots, n_k$ respectively, $N = \sum_{i=1}^{k} n_i$. In the genomic scenario, we may want to see the effect of k different tissue types (say liver, brain, heart, muscle, lungs, etc.) on the gene expression levels of a set of N genes, divided into k classes of sizes $n_1, n_2, \ldots, n_k$, respectively.

We use the following notations for the ANOVA:

$$\overline{X}_{i.} = \frac{\sum_{j=1}^{n_i} X_{ij}}{n_i}, \quad T_{i.} = \sum_{j=1}^{n_i} X_{ij},$$

$$G = T_{1.} + T_{2.} + \cdots + T_{k.} = \sum_{i=1}^{k} T_{i.} = \sum_{i=1}^{k} \sum_{j=1}^{n_i} X_{ij}.$$

In tabular form, we represent the one-way classification of data as shown in Table 9.1.1.

One-way ANOVA is not a matrix of size $k \times n_k$, since each treatment may have different sample sizes. If the sample size is the same (say n) for each treatment, then only it will be a matrix of size $k \times n$, which is common in genomic studies.

In one-way ANOVA the observations within a class (or treatment or condition) varies around its mean, and this type of variability is called "within group" variability. Also, the mean of each class varies around an overall mean of all the classes. This type of variability is known as "between classes" or "inter-group"

Table 9.1.1 One-Way Classified Data

Treatment	Observations				Means	Total
1	X_{11}	X_{12}	$\cdots$	X_{1n_1}	$\overline{X}_{1.}$	$T_{1.}$
2	X_{21}	X_{22}	$\cdots$	X_{2n_2}	$\overline{X}_{2.}$	$T_{2.}$
.	.	.		.	.	.
.	.	.		.	.	.
.	.	.		.	.	.
i	X_{i1}	X_{i2}	$\cdots$	X_{in_i}	$\overline{X}_{i.}$	$T_{i.}$
.	.	.		.	.	.
.	.	.		.	.	.
.	.	.		.	.	.
k	X_{k1}	X_{k2}	$\cdots$	X_{kn_k}	$\overline{X}_{k.}$	$T_{k.}$
						G

variability. Thus, an observation in the one-way ANOVA varies around its overall mean as well as its class mean. In ANOVA, we try to find the relationship between within-group and inter-group variabilities.

Let us consider the linear model, which is as follows:

$$X_{ij} = \mu + \alpha_i + \varepsilon_{ij}, \quad i = 1, 2, \ldots, k; j = 1, 2, \ldots, n_i. \qquad (9.1.1)$$

Here μ is the general mean effect given by

$$\mu = \frac{\sum_{i=1}^{k} n_i \mu_i}{N},$$

where μ_i is the fixed effect due to the ith treatment (condition or effect or tissue type etc.), α_i is the effect of the ith treatment given by

$$\alpha_i = \mu_i - \mu,$$

and ε_{ij} is the random error effect.

We make the following assumptions about the model (9.1.1):

1. All the observations X_{ij} are independent.

2. Effects are additive in nature.

3. The random effects are independently and identically distributed as $N(0, \sigma_\varepsilon^2)$.

We would like to test the equality of the population means. In other words, we test whether the mean effects of all the treatments are the same. We can state the null hypothesis as follows:

$$H_0 : \mu_1 = \mu_2 = \cdots = \mu_k = \mu, \tag{9.1.2}$$

or

$$H_0 : \alpha_1 = \alpha_2 = \cdots = \alpha_k = 0.$$

To estimate the parameters μ, and α_i in model (9.1.1), we use the principle of least squares on minimizing the error sum of squares. The error sum of squares is given by

$$E = \sum_{i=1}^{k} \sum_{j=1}^{n_i} \varepsilon_{ij}^2 = \sum_{i=1}^{k} \sum_{j=1}^{n_i} (X_{ij} - \mu - \alpha_i)^2.$$

Partially differentiating with respect to μ, and α_i, and equating $\frac{\partial E}{\partial \mu}$, and $\frac{\partial E}{\partial \alpha_i}$ equal to zero, we get

$$\hat{\mu} = \frac{\sum_{i=1}^{k} \sum_{j}^{n_i} X_{ij}}{N} = \overline{X}_{..}, \text{ using } \sum_{i=1}^{k} n_i \alpha_i = 0.$$

and

$$\hat{\alpha}_i = \sum_{j=1}^{n_i} \frac{X_{ij}}{n_j} - \hat{\mu} = \overline{X}_{i.} - \hat{\mu} = \overline{X}_{i.} - \overline{X}_{..}.$$

We find that we can write

$$\sum_{i=1}^{k} \sum_{j=1}^{n_i} (X_{ij} - \overline{X}_{..})^2 = \sum_{i=1}^{k} \sum_{j=1}^{n_i} (X_{ij} - \overline{X}_{i.} + \overline{X}_{i.} - \overline{X}_{..})^2$$

$$= \sum_{i=1}^{k} \sum_{j=1}^{n_i} (X_{ij} - \overline{X}_{i.})^2 + \sum_{i=1}^{k} n_i (\overline{X}_{i.} - \overline{X}_{..})^2 + 2 \left[\sum_{i=1}^{k} (\overline{X}_{i.} - \overline{X}_{..}) \sum_{j=1}^{n_i} (X_{ij} - \overline{X}_{i.}) \right].$$

$$\tag{9.1.3}$$

But

$$\sum_{j=1}^{n_i}(X_{ij}-\overline{X}_{i.})=0.$$

Therefore, (9.1.3) becomes

$$\sum_{i=1}^{k}\sum_{j=1}^{n_i}(X_{ij}-\overline{X}_{..})^2=\sum_{i=1}^{k}\sum_{j=1}^{n_i}(X_{ij}-\overline{X}_{i.})^2+\sum_{i=1}^{k}n_i(\overline{X}_{i.}-\overline{X}_{..})^2. \qquad (9.1.4)$$

Let

$$\text{Total Sum of Square (TSS)}=S_T^2=\sum_{i=1}^{k}\sum_{j=1}^{n_i}(X_{ij}-\overline{X}_{..})^2$$

$$\text{Sum of Squares of Errors (SSE)}=S_E^2=\sum_{i=1}^{k}\sum_{j=1}^{n_i}(X_{ij}-\overline{X}_{i.})^2$$

$$\text{Sum of Squares due to Treatment (SST)}=S_t^2=\sum_{i=1}^{k}n_i(\overline{X}_{i.}-\overline{X}_{..})^2$$

Then we can write (9.1.4) as

$$\text{TSS}=\text{SSE}+\text{SST}. \qquad (9.1.5)$$

The degrees of freedom for TSS are computed from using N quantities, and therefore, the degree of freedom is $(N-1)$. One degree of freedom is lost because of the linear constraint

$$\sum_{i=1}^{k}\sum_{j=1}^{n_i}(X_{ij}-\overline{X}_{..})=0.$$

The SST has $(k-1)$ degrees of freedom, since it is computed from k quantities, and one degree of freedom is lost because of the constraint

$$\sum_{i=1}^{k}n_i(\overline{X}_{i.}-\overline{X}_{..})=0.$$

The SSE has $(N-k)$ degrees of freedom since it is calculated from N quantities, which are subject to k constraints

$$\sum_{j=1}^{n_i}(X_{ij}-\overline{X}_{i.})=0, \quad i=1,2,\ldots,k.$$

The mean sum of squares (MSS) is obtained when sum of squares is divided by its degrees of freedom.

To test the null hypothesis, (9.1.2), we use the F-test given by

$$F = \frac{\text{MST}}{\text{MSE}} \sim F_{k-1,N-k}$$

where

$$\text{Mean sum of squares due to treatments} = \text{MST} = \frac{S_t^2}{k-1},$$

$$\text{Mean sum of squares due to error} = \text{MSE} = \frac{S_E^2}{N-k}.$$

If the null hypothesis is true, then the computed value of F must be close to 1; otherwise, it will be greater than 1, which will lead to rejection of the null hypothesis.

These results can be written using a table known as one-way analysis of variance (ANOVA) table as shown in Table 9.1.2.

ANOVA is called analysis of variance, but in a true sense, it is an analysis of a sum of squares.

To save computation time, we can use the following simplified versions of SST, SSE, and TSS:

$$\text{TSS} = S_T^2 = \sum_{i=1}^{k} \sum_{j=1}^{n_i} (X_{ij} - \overline{X}_{..})^2 = \sum_{i=1}^{k} \sum_{j=1}^{n_i} X_{ij}^2 - \frac{\left[\sum_{i=1}^{k} \sum_{j=1}^{n_i} X_{ij}\right]^2}{N}$$

$$= \sum_{i=1}^{k} \sum_{j=1}^{n_i} X_{ij}^2 - \frac{[G]^2}{N}. \tag{9.1.6}$$

Table 9.1.2 ANOVA Table for One-Way Classification

Source of Variations	Sum of Squares	Degrees of Freedom	Mean Sum of Squares	Variance Ratio Test
Treatment	S_t^2	$k-1$	$\text{MST} = \frac{S_t^2}{k-1}$	
Error	S_E^2	$N-k$	$\text{MSE} = \frac{S_E^2}{N-k}$	$F = \frac{\text{MST}}{\text{MSE}} \sim F_{k-1,N-k}$
Total	S_T^2	$N-1$		

$$SSE = S_E^2 = \sum_{i=1}^{k}\sum_{j=1}^{n_i}(X_{ij} - \overline{X}_{i.})^2 = \sum_{i=1}^{k}\left[\sum_{j=1}^{n_i}(X_{ij} - \overline{X}_{i.})^2\right]$$

$$= \sum_{i=1}^{k}\left[\sum_{j=1}^{n_i}X_{ij}^2 - \frac{\left(\sum_{j=1}^{n_i}X_{ij}\right)^2}{n_i}\right]$$

$$= \sum_{i=1}^{k}\sum_{j=1}^{n_i}X_{ij}^2 - \sum_{i=1}^{k}\frac{T_{i.}^2}{n_i}. \qquad (9.1.7)$$

Then we can find SST by using the relation

$$SST = TSS - SSE.$$

■ Example 1

Let a gene be suspected to have some connection with blood cancer. There are four stages of blood cancer: stage I, stage II, stage III, and stage IV. For treating a patient, identification of blood cancer is crucial in the first three stages. Three mRNA samples were collected from stage I, stage II, and stage III, respectively. The experiment is repeated six times, as shown in Table 9.1.3.

Find whether there is any difference in mean expression values in the three mRNA samples.

Solution

We set up the null hypothesis as follows:

$$H_0 : \mu_{\text{mRNA1}} = \mu_{\text{mRNA2}} = \mu_{\text{mRNA3}},$$

Table 9.1.3 mRNA Samples, One-Way Classification

	1	2	3	4	5	6
mRNA − 1	95	98	100	105	85	88
mRNA − 2	94	92	78	88	92	91
mRNA − 3	72	88	82	73	75	77

against

$$H_A : \mu_{mRNA1} \neq \mu_{mRNA2} \neq \mu_{mRNA3}.$$

We find that

	1	2	3	4	5	6	$T_{i.}$	$\sum\limits_{i=1}^{k} X_{ij}^2$
mRNA – 1	95	98	100	105	85	88	571	54623
mRNA – 2	94	92	78	88	92	91	535	47873
mRNA – 3	72	88	82	73	75	77	467	36535
						Total	1573	139031

The summary of the data is given in the following tables.

SUMMARY				
Groups	**Count**	**Sum**	**Average**	**Variance**
mRNA 1	6	571	95.16667	56.56667
mRNA 2	6	535	89.16667	33.76667
mRNA 3	6	467	77.83333	37.36667

ANOVA						
Source of Variation	*SS*	*df*	*MS*	*F*	**P-value**	*F* **critical**
Between Groups (Treatment)	929.7778	2	464.8889	10.92143	0.001183	3.68232
Within Groups (Error)	638.5000	15	42.5666			
Total	1568.2780	17				

Since the p-value is $0.001183 < \alpha = 0.05$, we reject the claim that the mean values of the mRNA samples are equal. Therefore, based on the given sample, we conclude that three stages have different mean expression values.

Following is the R-code for example 1.

```
> # Example 1
> rm(list = ls( ))
>
> mrna <- data.frame(expr = c(95, 98, 100, 105, 85, 88,94, 92, 78, 88, 92, 91, 72,
88, 82, 73, 75, 77),
+ samples = factor (c(rep ("mRNA1",6), rep("mRNA2", 6), rep("mRNA3", 6))))
> mrna
expr samples
  1    95  mRNA1
  2    98  mRNA1
  3   100  mRNA1
  4   105  mRNA1
  5    85  mRNA1
  6    88  mRNA1
  7    94  mRNA2
  8    92  mRNA2
  9    78  mRNA2
 10    88  mRNA2
 11    92  mRNA2
 12    91  mRNA2
 13    72  mRNA3
 14    88  mRNA3
 15    82  mRNA3
 16    73  mRNA3
 17    75  mRNA3
 18    77  mRNA3
>
> # mean, variance and standard deviation
>
> sapply(split(mrna$expr,mrna$samples),mean)
mRNA1   mRNA2   mRNA3
95.16667   89.16667   77.83333
>
> sapply(split(mrna$expr,mrna$samples,var)
mRNA1   mRNA2   mRNA3
56.56667   33.76667   37.36667
>
> sqrt(sapply(split(mrna$expr,mrna$samples), var))
mRNA1   mRNA2   mRNA3
7.521081   5.810909   6.112828
>
> # Box-Plot
> boxplot(split(mrna$expr,mrna$samples), xlab = "Samples",ylab = "Expression
```

```
Levels", col = "red")
> # See Figure 9.1.1
> # lm( ) to fit a Linear Model
>
> fitmrna<- lm(expr~samples, data = mrna)
>
> #Displays the result
>
> summary(fitmrna)
Call:
lm(formula = expr ~ samples, data = mrna)
Residuals:
Min   1Q   Median   3Q   Max
-11.1667   -4.3333   0.8333   3.8333   10.1667
Coefficients:
Estimate Std. Error t value Pr(>|t|)
(Intercept)   95.167   2.664   35.729   6.25e-16***
samplesmRNA2   -6.000   3.767   -1.593   0.132042
samplesmRNA3   -17.333   3.767   -4.602   0.000346***
---
Signif. codes: 0'***'0.001 '**'0.01'*'0.05'.'0.1 '1
Residual standard error: 6.524 on 15 degrees of freedom
Multiple R-squared: 0.5929, Adjusted R-squared: 0.5386
F-statistic: 10.92 on 2 and 15 DF, p-value: 0.001183
> # Use anova( )
>
> anova(fitmrna)
Analysis of Variance Table
Response: expr
Df Sum Sq Mean Sq F value Pr(>F)
samples 2 929.78 464.89 10.921 0.001183 **
Residuals 15 638.50 42.57
---
Signif. codes: 0'***' 0.001 '**'0.01'*'0.05'.' 0.1' '1
>
```

■ Example 2

A group of mice was inoculated with five strains of malaria organisms to observe the number of days each mouse survived so that a treatment strategy could be developed. Table 9.1.4 gives the number of days each mouse survived. Six mice were inoculated with each strain. Find whether the mean effect of different strains of malaria organisms is the same.

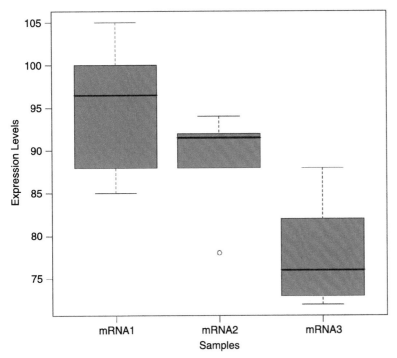

FIGURE 9.1.1

Box plot for expression levels.

Table 9.1.4 Number of Days Each Mouse Survived					
	1	**2**	**3**	**4**	**5**	**6**
Strain A	10	9	8	11	12	10
Strain B	6	7	8	6	5	5
Strain C	8	7	9	10	9	9
Strain D	11	10	12	13	11	12
Strain E	6	5	7	6	5	5

Solution

The null hypothesis to be tested is as follows:

$$H_0 : \mu_A = \mu_B = \mu_C = \mu_D = \mu_E$$

against

$$H_1 : \mu_A \neq \mu_B \neq \mu_C \neq \mu_D \neq \mu_E.$$

We compute $T_{i.}$ and $\sum_{i=1}^{k} X_{ij}^2$.

	1	2	3	4	5	6	$T_{i.}$	$\sum_{i=1}^{k} X_{ij}^2$
Strain A	10	9	8	11	12	10	60	610
Strain B	6	7	8	6	5	5	37	235
Strain C	8	7	9	10	9	9	52	456
Strain D	11	10	12	13	11	12	69	799
Strain E	6	5	7	6	5	5	34	196
						Total	252	2296

A one-way classifed ANOVA is performed, and the results are given in the following tables:

SUMMARY				
Groups	**Count**	**Sum**	**Average**	**Variance**
Strain 1	6	60	10	2
Strain 2	6	37	6.166667	1.366667
Strain 3	6	52	8.666667	1.066667
Strain 4	6	69	11.5	1.1
Strain 5	6	34	5.666667	0.666667

ANOVA						
Source of Variation	*SS*	*df*	*MS*	*F*	*P*-value	*F* critical
Between Groups (Treatment)	148.2	4	37.05	29.87903	33.87E-09	2.75871
Within Groups (Error)	31	25	1.24			
Total	179.2	29				

Since the p-value is $3.387 \times 10^{-9} < \alpha = 0.05$, we reject the null hypothesis that mean effects of the different strains of malaria are the same.

The R-code for example 2 is as follows:

```
> #Example 2
> rm(list = ls())
>
> strain <-data.frame(days = c(10,9,8,11,12,10,
+ 6,7,8,6,5,5,
+ 8,7,9,10,9,9,
+ 11,10,12,13,11,12,
+ 6,5,7,6,5,5),
```

```
+ numberofdays = factor(c(rep("StrainA",6), rep("StrainB", 6),
+ rep("StrainC", 6),rep("StrainD", 6),rep("StrainE", 6))))
> strain
days numberofdays
 1   10   StrainA
 2    9   StrainA
 3    8   StrainA
 4   11   StrainA
 5   12   StrainA
 6   10   StrainA
 7    6   StrainB
 8    7   StrainB
 9    8   StrainB
10    6   StrainB
11    5   StrainB
12    5   StrainB
13    8   StrainC
14    7   StrainC
15    9   StrainC
16   10   StrainC
17    9   StrainC
18    9   StrainC
19   11   StrainD
20   10   StrainD
21   12   StrainD
22   13   StrainD
23   11   StrainD
24   12   StrainD
25    6   StrainE
26    5   StrainE
27    7   StrainE
28    6   StrainE
29    5   StrainE
30    5   StrainE
>
> # mean, variance and standard deviation
>
> sapply(split(strain$days,strain$numberofdays),mean)
StrainA   StrainB   StrainC   StrainD   StrainE
10.000000  6.166667  8.666667  11.500000  5.666667
>
> sapply(split(strain$days,strain$numberofdays),var)
StrainA   StrainB   StrainC   StrainD   StrainE
2.0000000  1.3666667  1.0666667  1.1000000  0.6666667
>
> sqrt(sapply(split(strain$days,strain$numberofdays),var))
```

```
StrainA   StrainB   StrainC   StrainD   StrainE
1.4142136   1.1690452   1.0327956   1.0488088   0.8164966
>
> #Box-Plot
>
> boxplot(split(strain$days,strain$numberofdays),xlab = "Strain",ylab = "Number of
Days",col = "blue")
> # See Figure 9.1.2
> # lm() to fit a Linear Model
>
> fitstrain<- lm(days~numberofdays, data=strain)
>
> #Displays the result
>
> summary(fitstrain)
Call:
lm(formula = days ~ numberofdays, data = strain)
Residuals:
Min 1Q Median 3Q Max
-2.000e+00 -6.667e-01 -1.153e-15 5.000e-01 2.000e+00
Coefficients:
Estimate Std. Error t value Pr(>|t|)
(Intercept) 10.0000 0.4546 21.997 < 2e-16***
numberofdaysStrainB -3.8333 0.6429 -5.962 3.17e-06***
numberofdaysStrainC -1.3333 0.6429 -2.074 0.0485*
numberofdaysStrainD 1.5000 0.6429 2.333 0.0280*
numberofdaysStrainE -4.3333 0.6429 -6.740 4.60e-07***
---
Signif. codes: 0 '***' 0.001 '**' 0.01'*' 0.05'.' 0.1 '' 1
Residual standard error: 1.114 on 25 degrees of freedom
Multiple R-squared: 0.827, Adjusted R-squared: 0.7993
F-statistic: 29.88 on 4 and 25 DF, p-value: 3.387e-09
>
> #Use anova()
>
> anova(fitstrain)
Analysis of Variance Table
Response: days
Df   SumSq   Mean Sq F value   Pr(>F)
numberofdays 4 148.20   37.05   29.879 3.387e-09***
Residuals   25   31.00   1.24
---
Signif. codes: 0 '***' 0.001 '**' 0.01 '*' 0.05 '.' 0.1 '' 1
>
>
```

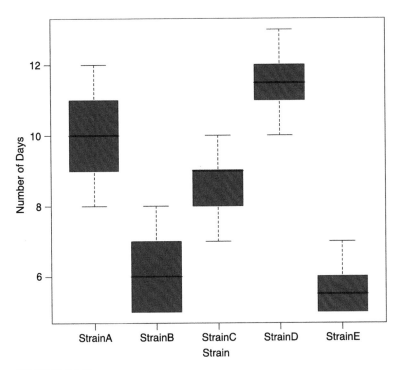

FIGURE 9.1.2

Box plot for strains.

Sometimes we are interested in finding whether there is a significant variability among different treatments. The data set is put in a similar format as shown in Table 9.1.1. The null hypothesis takes the form

$$H_0 : \sigma^2_{\text{Treatments}} = 0, \qquad (9.1.8)$$

against

$$H_1 : \sigma^2_{\text{Treatments}} > 0.$$

An ANOVA table similar to Table 9.1.2 is constructed to test the null hypothesis given in (9.1.8).

9.2 TWO-WAY CLASSIFICATION OF ANOVA

In many situations, the data might be influenced by more than one factor. For example, in the case of sex-linked recessive lethal mutation in *Drosophila*, the mutation can be induced by ionizing radiations such as X-ray as well as by chemical substances that interact with DNA to create base changes.

Let us consider that there are N observations $X_{ij}, i = 1, 2, \ldots, k; j = 1, 2, \ldots, h$ of a random variable X grouped (known as treatments or effects or conditions) into k classes, and each class (sometimes called a block) is of size h; thus, $N = k \times h$. If we look at the genomic situation, we consider h mRNA samples hybridized on k arrays. Each expression value can be classified according to the mRNA sample to which it belongs or according to the array (or blocks) from which it is generated.

We define the following notations for the ANOVA:

$$\overline{X}_{i.} = \frac{\sum_{j=1}^{h} X_{ij}}{h}, \overline{X}_{.j} = \frac{\sum_{i=1}^{k} X_{ij}}{k} T_{i.} = \sum_{j=1}^{h} X_{ij}, T_{.j} = \sum_{i=1}^{k} X_{ij},$$

$$G = T_{1.} + T_{2.} + \cdots + T_{k.} = \sum_{i=1}^{k} T_{i.} = \sum_{j=1}^{h} T_{.j} = \sum_{i=1}^{k} \sum_{j=1}^{h} X_{ij}.$$

In tabular form, we represent the two-way classification of data as shown in Table 9.2.1.

In two-way ANOVA, the observations in a treatmentwise classifed class vary around its mean, which is called "variability due to treatment," and it also varies according to blockwise classification around its mean, which is called "variability due to block." Also, the mean of each class varies around an overall

Table 9.2.1 Two-Way Classified Data

		Classification-2 (Blocks)				Means	Total
		1	2	$\cdots$	h		
Classification-1	1	X_{11}	X_{12}	$\cdots$	X_{1h}	$\overline{X}_{1.}$	$T_{1.}$
(Treatments)	2	X_{21}	X_{22}	$\cdots$	X_{2h}	$\overline{X}_{2.}$	$T_{2.}$
	.	.	.		.	.	.
	.	.	.		.	.	.
	i	X_{i1}	X_{i2}	$\cdots$	X_{ih}	$\overline{X}_{i.}$	$T_{i.}$
	.	.	.		.	.	.
	.	.	.		.	.	.
	k	X_{k1}	X_{k2}	$\cdots$	X_{kh}	$\overline{X}_{k.}$	$T_{k.}$
Mean		$\overline{X}_{.1}$	$\overline{X}_{.2}$	$\cdots$	$\overline{X}_{.h}$	$\overline{X}_{..}$	
Total		$T_{.1}$	$T_{.2}$	$\cdots$	$T_{.h}$		G

mean of all the classes. Therefore, we find that an observation in the two-way ANOVA varies around its overall mean as well as its treatment and block means.

For two-way ANOVA, we consider the linear model

$$X_{ij} = \mu + \alpha_i + \beta_j + \gamma_{ij} + \varepsilon_{ij}, \quad i = 1, 2, \ldots, k; j = 1, 2, \ldots, h. \quad (9.2.1)$$

Here we define μ as the general mean effect given by

$$\mu = \frac{\sum_{i=1}^{k} \sum_{j=1}^{h} \mu_{ij}}{N}$$

where μ_{ij} is the fixed effect due to ith treatment (condition or effect or tissue type, etc.) and jth block.

α_i is the effect of the ith treatment given by

$$\alpha_i = \mu_{i.} - \mu, \text{ with } \mu_{i.} = \frac{\sum_{j=1}^{h} \mu_{ij}}{h},$$

β_j is the effect of the jth block given by

$$\beta_j = \mu_{.j} - \mu, \text{ with } \mu_{.j} = \frac{\sum_{i=1}^{k} \mu_{ij}}{k},$$

The interaction effect γ_{ij} due to the ith treatment and jth block is given by

$$\gamma_{ij} = \mu_{ij} - \mu_{i.} - \mu_{.j} + \mu$$

and ε_{ij} is the random error effect.

We make the following assumptions about the model (9.2.1):

1. All the observations X_{ij} are independent.

2. Effects are additive in nature.

3. The random effects are independently and identically distributed as $N\left(0, \sigma_\varepsilon^2\right)$.

We observe that there is only one observation per cell, the observation corresponding to the ith treatment and jth block, denoted by X_{ij}. The interaction effect γ_{ij} cannot be estimated by using only one observation; therefore, the interaction effect $\gamma_{ij} = 0$ and the model (9.2.1) reduce to

$$X_{ij} = \mu + \alpha_i + \beta_j + \varepsilon_{ij}, \quad i = 1, 2, \ldots, k; j = 1, 2, \ldots, h. \quad (9.2.2)$$

We would like to test the equality of the population means—that is, that mean effects of all the treatments as well as blocks are the same. We state it as follows:

$$H_0 : \begin{cases} H_t : \mu_{1.} = \mu_{2.} = \cdots = \mu_{k.} = \mu. \\ H_b : \mu_{.1} = \mu_{.2} = \cdots = \mu_{.k} = \mu. \end{cases} \tag{9.2.3}$$

or

$$H_0 : \begin{cases} H_t : \alpha_1 = \alpha_2 = \cdots = \alpha_k = 0. \\ H_b : \beta_1 = \beta_2 = \cdots = \beta_k = 0. \end{cases} \tag{9.2.4}$$

To estimate the parameters μ, α_i, and β_j in model (9.2.2), we use the principle of least squares on minimizing the error sum of squares. The error sum of squares is given by

$$E = \sum_{i=1}^{k}\sum_{j=1}^{h} \varepsilon_{ij}^2 = \sum_{i=1}^{k}\sum_{j=1}^{h}(X_{ij} - \mu - \alpha_i - \beta_j)^2.$$

Partially differentiating with respect to μ, α_i, and β_j, and equating $\frac{\partial E}{\partial \mu}, \frac{\partial E}{\partial \alpha_i}$, and $\frac{\partial E}{\partial \beta_j}$ equal to zero, we get

$$\hat{\mu} = \frac{\sum_{i=1}^{k}\sum_{j}^{h} X_{ij}}{N} = \overline{X}_{..},$$

$$\hat{\alpha}_i = \sum_{j=1}^{h} \frac{X_{ij}}{h} - \hat{\mu} = \overline{X}_{i.} - \hat{\mu} = \overline{X}_{i.} - \overline{X}_{..},$$

$$\hat{\beta}_j = \sum_{i=1}^{k} \frac{X_{ij}}{k} - \hat{\mu} = \overline{X}_{.j} - \hat{\mu} = \overline{X}_{.j} - \overline{X}_{..},$$

using $\sum_{i=1}^{k} \alpha_i = 0, \sum_{j=1}^{h} \beta_j = 0$.

Using the fact that the algebraic sum of deviations of measurements from their mean is zero and product terms are zero, we write

$$\sum_{i=1}^{k}\sum_{j=1}^{h}(X_{ij} - \overline{X}_{..})^2 = \sum_{i=1}^{k}\sum_{j=1}^{n_i}(X_{ij} - \overline{X}_{i.} + \overline{X}_{i.} - \overline{X}_{.j} + \overline{X}_{.j} - \overline{X}_{..})^2$$

$$= h\sum_{i=1}^{k}(\overline{X}_{i.} - \overline{X}_{..})^2 + k\sum_{j=1}^{h}(\overline{X}_{.j} - \overline{X}_{..})^2 + \sum_{i=1}^{k}\sum_{j=1}^{h}(X_{ij} - \overline{X}_{i.} - \overline{X}_{.j} - \overline{X}_{..})^2. \tag{9.2.5}$$

Let

$$\text{Total Sum of Square (TSS)} = S_T^2 = \sum_{i=1}^{k}\sum_{j=1}^{h}(X_{ij} - \overline{X}_{..})^2,$$

$$\text{Sum of Squares of Errors (SSE)} = S_E^2 = \sum_{i=1}^{k}\sum_{j=1}^{h}(X_{ij} - \overline{X}_{i.} - \overline{X}_{.j} - \overline{X}_{..})^2,$$

$$\text{Sum of Squares due to Treatment (SST)} = S_t^2 = h\sum_{i=1}^{k}(\overline{X}_{i.} - \overline{X}_{..})^2,$$

$$\text{Sum of Squares due to blocks (SSB)} = S_b^2 = k\sum_{j=1}^{h}(\overline{X}_{.j} - \overline{X}_{..})^2.$$

Then we write (9.2.5) as

$$S_T^2 = S_t^2 + S_b^2 + S_E^2 \tag{9.2.6}$$

or

$$\text{TSS} = \text{SST} + \text{SSB} + \text{SSE}.$$

The degrees of freedom for TSS are computed from using N quantities, and therefore, the degree of freedom is $(N - 1)$; one degree of freedom is lost because of the linear constraint

$$\sum_{i=1}^{k}\sum_{j=1}^{n_i}(X_{ij} - \overline{X}_{..}) = 0.$$

The SST has $(k - 1)$ degrees of freedom, since it is computed from k quantities, and one degree of freedom is lost because of the constraint

$$\sum_{i=1}^{k}(\overline{X}_{i.} - \overline{X}_{..}) = 0.$$

The SSB has $(h - 1)$ degrees of freedom, since it is computed from h quantities, and one degree of freedom is lost because of the constraint

$$\sum_{j=1}^{h}(\overline{X}_{.j} - \overline{X}_{..}) = 0.$$

The SSE has $(N - 1) - (k - 1) - (h - 1) = (h - 1)(k - 1)$ degrees of freedom.

The mean sum of squares (MSS) is obtained when the sum of squares is divided by its degrees of freedom.

To test the null hypothesis, H_t (9.2.3), we use the F-test given by

$$F = \frac{\text{MST}}{\text{MSE}} \sim F_{k-1,(k-1)(h-1)}$$

and to test the null hypothesis, H_b (9.2.3), we use the F-test given by

$$F = \frac{\text{MSB}}{\text{MSE}} \sim F_{h-1,(k-1)(h-1)}$$

where

$$\text{Mean sum of squares due to treatments} = \text{MST} = \frac{S_t^2}{k-1},$$

$$\text{Mean sum of squares due to blocks (varities)} = \text{MSB} = \frac{S_b^2}{h-1},$$

$$\text{Mean sum of squares due to errors} = \text{MSE} = \frac{S_E^2}{(k-1)(h-1)}.$$

If the null hypothesis is true, then the computed value of F must be close to 1; otherwise, it will be greater than 1, which will lead to rejection of the null hypothesis.

These results are written using a table known as two-way analysis of variance (ANOVA) as shown in Table 9.2.2.

The two-way classified data with one observation per cell does not allow us to estimate the interaction effect between two factors. For this, we need more than one observation per cell—say m observations per cell. If we have m observations

Table 9.2.2 ANOVA Table for Two-Way Classification

Source of Variations	Sum of Squares	Degrees of Freedom	Mean Sum of Squares	Variance Ratio
Treatment	S_t^2	$k-1$	$\text{MST} = \frac{S_t^2}{k-1}$	$F = \frac{\text{MST}}{\text{MSE}} \sim F_{k-1,(k-1)(h-1)}$
Blocks	S_b^2	$h-1$	$\text{MSB} = \frac{S_b^2}{h-1}$	$F = \frac{\text{MSB}}{\text{MSE}} \sim F_{h-1,(k-1)(h-1)}$
Error	S_E^2	$(k-1)(h-1)$	$\text{MSE} = \frac{S_E^2}{(k-1)(h-1)}$	
Total	S_T^2	$N-1$		

per cell, then the ANOVA is called two-way classified data with m observations per cell.

■ Example 1

Effect of Atorvastatin (Lipitor) on Gene Expression in People with Vascular Disease (National Institutes of Health)

It has been known that atherosclerosis and its consequences—coronary heart disease and stroke—are the principal causes of mortality. Gene expression profiling of peripheral white blood cells provides information that may be predictive about vascular risk. Table 9.2.3 gives gene expression measurements, classified according to age group and dose level of Atorvastatin treatment.

Test whether the dose level and age groups significantly affect the gene expression.

Solution

Using Table 9.2.2, we get the following results:

ANOVA: Two-Factor with one observation per cell

ANOVA

For the classifications according to the age groups, we find that since $p = 0.013562 < \alpha = 0.05$; therefore, we reject the claim that there is no significant effect of age group on gene expression measurements. For the classification according to dose level, we find that $p = 0.169122 > \alpha = 0.05$;

Table 9.2.3 Gene Expression Data for Atherosclerosis

		Dose Level			
		1	2	3	4
	1	100	99	87	98
	2	95	94	83	92
	3	102	80	86	85
Age Group	4	84	82	80	83
	5	77	78	90	76
	6	90	76	74	85

			Dose level			Means	Total
		1	2	3	4		
Age group	1	100	99	87	98	96	384
	2	95	94	83	92	91	364
	3	102	80	86	85	88.25	353
	4	84	82	80	83	82.25	329
	5	77	78	90	76	80.25	321
	6	90	76	74	85	81.25	325
Means		91.33	84.83	83.33	86.5		
Totals		548	509	500	519		2076

SUMMARY	Count	Sum	Average	Variance
Age group-1	4	384	96	36.66666667
Age group-2	4	364	91	30
Age group-3	4	353	88.25	90.91666667
Age group-4	4	329	82.25	2.916666667
Age group-5	4	321	80.25	42.91666667
Age group-6	4	325	81.25	56.91666667
Dose level-1	6	548	91.33333	92.66666667
Dose level-2	6	509	84.83333	88.16666667
Dose level-3	6	500	83.33333	32.66666667
Dose level-4	6	519	86.5	57.9

Source of Variation	SS	df	MS	F	P-value	F critical
Age group	793	5	158.6	4.218085106	0.013562	2.901295
Dose level	217	3	72.33333	1.923758865	0.169122	3.287382
Error	564	15	37.6			
Total	1574	23				

we fail to reject the claim that there is no significant effect due to dose level on gene expression measurements.

The following R-code performs ANOVA for example 1.

```
> #Example 1
> rm(list = ls())
>
> effect <- data. frame(geneexp = c(100,99,87,98,
```

```
+ 95, 94, 83, 92,
+ 102, 80, 86, 85,
+ 84, 82, 80, 83,
+ 77, 78, 90, 76,
+ 90, 76, 74, 85),
+ agegroup = factor (rep(rep(1:6, rep(4,6)),1))
+ ,dose = factor (rep(c("D1","D2","D3","D4"), c(1,1,1,1))))
> effect
geneexp agegroup dose
  1   100   1 D1
  2    99   1 D2
  3    87   1 D3
  4    98   1 D4
  5    95   2 D1
  6    94   2 D2
  7    83   2 D3
  8    92   2 D4
  9   102   3 D1
 10    80   3 D2
 11    86   3 D3
 12    85   3 D4
 13    84   4 D1
 14    82   4 D2
 15    80   4 D3
 16    83   4 D4
 17    77   5 D1
 18    78   5 D2
 19    90   5 D3
 20    76   5 D4
 21    90   6 D1
 22    76   6 D2
 23    74   6 D3
 24    85   6 D4
>
>
> # mean, variance and standard deviation
>
> sapply(splitt(effect$geneexp, effect$agegroup), mean)
   1     2     3     4     5     6
96.00 91.00 88.25 82.25 80.25 81.25
>
> sapply(splitt(effect$geneexp, effect$dose), mean)
  D1       D2       D3       D4
91.33333 84.83333 83.33333 86.50000
>
```

```
> sapply(splitt(effect$geneexp,effect$agegroup), var)
1   2   3   4   5   6
36.666667   30.000000   90.916667   2.916667   42.916667   56.916667
>
> sapply(splitt(effect$geneexp,effect$dose), var)
D1   D2   D3   D4
92.66667   88.16667   32.66667   57.90000
>
> sqrt(sapply(splitt(effect$geneexp,effect$agegroup), var))
1   2   3   4   5   6
6.055301   5.477226   9.535023   1.707825   6.551081   7.544314
>
> sqrt(sapply(splitt(effect$geneexp,effect$dose), var))
D1   D2   D3   D4
9.626353   9.389711   5.715476   7.609205
>
>
> #Box-Plot
>
> boxplot(split(effect$geneexp,effect$agegroup),xlab = "Age Group", ylab = "Gene
Expression",col = "blue")
> boxplot(split(effect$geneexp,effect$dose),xlab = "Dose",ylab = "Gene
Expression",col = "blue")
> # See Figure 9.2.1, and Figure 9.2.2
> fitexp<-lm(geneexp~agegroup+dose, data = effect)
> #Display Results
> fitexp
Call:
lm(formula = geneexp ~ agegroup + dose, data = effect)
Coefficients:
(Intercept)  agegroup2  agegroup3  agegroup4  agegroup5  agegroup6
100.833   -5.000   -7.750   -13.750   -15.750   -14.750
doseD2   doseD3   doseD4
-6.500   -8.000   -4.833
>> #Use anova()
>
> anova(fitexp)
Analysis of Variance Table
Response: geneexp
         Df  Sum Sq  Mean Sq  F value  Pr(>F)
agegroup  5   793.00  158.60   4.2181   0.01356*
dose      3   217.00   72.33   1.9238   0.16912
Residuals 15  564.00   37.60
Signif. codes: 0 '***' 0.001 '**' 0.01 '*' 0.05 '.' 0.1 ' ' 1
>
```

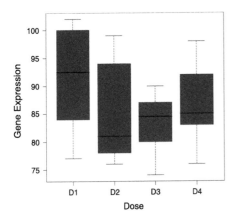

FIGURE 9.2.1

Box plot for gene expression and dose levels.

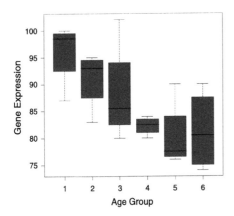

FIGURE 9.2.2

Box plot for gene expression and age group.

The ANOVA method is used in the design of experiments, which is presented in Chapter 10. ∎

The Design of Experiments

CONTENTS

10.1 INTRODUCTION

The design of experiments in genomics studies, in particular microarray experiments, has been used extensively to provide a measure of different developmental stages or experimental conditions. The design of experiments allows researchers to balance different considerations such as cost and accuracy. Many different designs will be available for the same scientific problem. The choice of design will depend on resources, cost, and desired level of accuracy. Microarray data are expensive to collect, and there are many sources of variations. There can be biological variations, technical variations, and measurement errors (Churchill, 2002). Biological variations are intrinsic to all organisms depending on genetic or environmental conditions. Technical variations can occur at different stages of experiments, such as extraction, labeling, and hybridization. Measurement errors may occur while converting signals to numerical values. The task of a statistical design is to reduce or eliminate the effect of unwanted variations and increase the precision of the quantities involved.

A typical experiment design involves

(i) Planning the experiment,

(ii) Obtaining information for testing the statistical hypothesis under study,

(iii) Performing statistical analysis of the data.

To obtain the desired information about the population characteristics from the sample, one needs to design the experiment in advance. The information obtained from an experiment that is carefully planned and well designed in advance gives entirely valid inferences.

There are two major categories of experiments: absolute and comparative. In absolute experiments, we are interested in determining the absolute value of some characteristics—for example, the average value of expression levels and the correlation between different gene expression measurements obtained from a set of genes. The comparative experiments are designed to compare the effect of two or more factors on some population characteristic—for example, the comparison of different treatments. Various objects of comparison in a comparative experiment are known as treatments—for example, in the case of microarray experiments, different dyes or different treatment conditions.

An experimental unit is the smallest division of the experimental material to which we apply treatments, and observations are made on the experimental unit of the variable under study. In agricultural experiments, the lot of "land" is the experimental unit. In genomic studies, the experimental unit may be patients, animals, or organisms under consideration. A block is a subset of experimental units that are expected to be homogenous with respect to some characteristics of the population under study. In agricultural experiments, often an experimental unit is divided into relatively homogeneous subgroups or strata. In genomic studies, blocking is used to reduce the variability due to differences between blocks. In microarray experiments, blocking is done by treating the microarray slide with the same processing technique such as hybridization, drying, etc. This way, all the spots on the same slide are subject to same processing technique; therefore, the intensity of light of fluorescent dye from the spots will be homogenous when excited by the laser. The errors, which are due to the nature of the experimental material and cannot be controlled, are known as experimental errors.

Replication means that we repeat an experiment more than once. There are two types of replications in genomic studies. One is known as a technical replicate, and the other is known as a biological replicate. The technical replication occurs when the same cell line is used to generate samples. The technical replication is considered when an inference is needed on only a single cell line under consideration. The biological replication occurs when biologically independent samples are taken from different cell lines. The biological replication is labor intensive and time consuming, but it is essential when a researcher is trying to make inferences about a biological condition. The number of replications, r, are directly linked to the amount of information contained in the design or precision, and it is given by

$$\frac{1}{Var(\bar{x})} = \frac{r}{\sigma^2} \qquad (10.1.1)$$

where σ^2 is the error variance per unit.

An efficient or sensitive design has a greater ability or power to detect the differences between treatment effects. From (10.1.1), we see that the efficiency of a design can be increased by decreasing σ^2, and by increasing r, the number of replications.

How many replicates one should choose has always been a question in genomic studies. Wernisch (2002) derived an approximate formula for the number of replicates, r, needed to achieve a $(1-\alpha)$ level of significance to detect f-fold differential expressions in a microarray experiment, which is given by

$$r \geq \frac{\sigma^2 \log f}{z_{1-\alpha}^2} \qquad (10.1.2)$$

where z is a standard normal variate, and $z_{1-\alpha}$ is the $(1-\alpha)$ quantile of the standard normal variate.

10.2 PRINCIPLES OF THE DESIGN OF EXPERIMENTS

There are three basic principles of the design of experiments:

(i) Replication,

(ii) Randomization,

(iii) Local control.

As discussed in section 10.1, replication means "the repetition of the treatment under investigation." Replication helps to reduce or eliminate the influence of chance factors on different experimental units. Thus, replication gives a more reliable estimate than the estimate that is obtained by using a single observation, and it also gives more confidence to researchers regarding the inference about population parameters. In many experimental situations, a general rule is to get as many replications as possible that should provide at least 12 degrees of freedom for the error. This follows from the fact that the values of the F-statistic do not decrease rapidly beyond $v_2 = 12$. Usually, one should not use fewer than four replications. In the case of microarray experiments, one can use the formula given in (10.1.2) to get the required number of replicates.

Randomization is a process of assigning treatments to various experimental units in a purely chance manner. This process ensures that no preferential allocation of treatment to some experimental units is made. Moreover, the validity of the statistical tests of significance (for example, the t-test for testing the significance of the difference of two means or the analysis of variance F-test for testing the homogeneity of several means) depends on the fact that the statistic under consideration follows the principle of randomization. The assumption of randomness is necessary because Standard Error (S.E.) of mean, given by S.E. $(\bar{x}) = \frac{\sigma}{\sqrt{n}}$ is valid only under random sampling. Randomization ensures that the sources of variation, which cannot be controlled in an experiment, operate randomly so that the average effect on any group of units is zero. In genomic studies, the biological samples such as mRNA samples from different tissues are generally selected randomly. For example, if we need to compare control and experimental subjects, we can divide the subjects into two groups randomly and then take the samples from these two groups.

The local control is the process by which the experimental error is reduced or eliminated. This process involves dividing the relatively heterogeneous experimental area into a homogeneous experimental area. This effect can be achieved through the use of several techniques that can divide the experimental material into homogenous groups (blocks) row-wise or column-wise (randomized block design) or both (Latin square design), according to the factors under consideration.

Some of the common designs available in the literature that are useful in genomic studies are discussed in the following sections.

10.3 COMPLETELY RANDOMIZED DESIGN

The simplest of all designs is a design known as completely randomized design (CRD). It uses the principles of randomization and replication. The treatments are allocated at random to the experimental units over the entire experimental material and replicated many times. Let there be k treatments, the ith treatment being replicated n_i times, $i = 1, 2, \ldots, k$. The experimental material is divided into $N = \sum n_i$ experimental units, and the treatments are applied randomly to the units with the condition that the ith treatment occurs n_i times. The CRD uses the entire experimental material and is very flexible because it does not complicate statistical analysis even if the replication of treatments is done an unequal number of times for each treatment. Also, a missing or lost observation does not complicate the analysis process. If the experimental material is not homogeneous in nature, then CRD is less informative than some of the other designs available. The CRD is more useful in laboratory settings than agricultural field settings because in laboratory settings it is possible to have homogenous experimental material or reduce the variability between experimental units. Biological variability can be controlled by keeping the same cell line, and the technical variability can be reduced by using the same technique and same machine for all the samples at once. The CRD works very well in the cases in which an appreciable fraction of experimental units is likely to be destroyed or to fail to respond during the treatment process. In the case of microarray experiments, after treatment, some of the spots fail to respond to laser scanning; therefore, it may be a good idea to use CRD for the statistical analysis of the data.

The statistical analysis of CRD is similar to the ANOVA for one-way classified data. The linear model is given by

$$y_{ij} = \mu + \tau_i + \varepsilon_{ij}, \quad i = 1, 2, \ldots, k; j = 1, 2, \ldots, n_i \qquad (10.3.1)$$

where y_{ij} is the response from the jth experimental unit receiving the ith treatment, μ is the general mean effect, τ_i is the effect due to the ith treatment, ε_{ij} is the error effect due to chance such that ε_{ij} is identically and independently distributed as $N(0, \sigma_e^2)$, and effects are assumed to be additive in nature. The $N = \sum_{i=1}^{k} n_i$ is the total number of experimental units. Using the same notations and analysis as in the case of one-way classified ANOVA, we get the CRD as shown in Table 10.3.1.

We would like to test the equality of the population means, that is, whether the mean effects of all the treatments are the same. We state this test as follows:

$$H_0 : \mu_1 = \mu_2 = \ldots = \mu_k = \mu, \qquad (10.3.2)$$

or

$$H_0 : \tau_1 = \tau_2 = \ldots = \tau_k = 0.$$

Table 10.3.1 CRD Data

Treatment	Replications				Means	Total
1	X_{11}	X_{12}	$\ldots$	X_{1n_1}	$\overline{X}_{1.}$	$T_{1.}$
2	X_{21}	X_{22}	$\ldots$	X_{2n_2}	$\overline{X}_{2.}$	$T_{2.}$
$\vdots$	$\vdots$	$\vdots$	$\vdots$	$\vdots$	$\vdots$	$\vdots$
i	X_{i1}	X_{i2}	$\ldots$	X_{in_i}	$\overline{X}_{i.}$	$T_{i.}$
$\vdots$	$\vdots$	$\vdots$	$\vdots$	$\vdots$	$\vdots$	$\vdots$
k	X_{k1}	X_{k2}	$\ldots$	X_{kn_k}	$\overline{X}_{k.}$	$T_{k.}$
						G

Table 10.3.2 CRD

Source of Variations	Sum of Squares	Degrees of Freedom	Mean Sum of Squares	Variance Ratio Test
Treatment	S_t^2	$k-1$	$\text{MST} = \frac{S_t^2}{k-1}$	
Error	S_E^2	$N-k$	$\text{MSE} = \frac{S_E^2}{N-k}$	$F = \frac{\text{MST}}{\text{MSE}} \sim F_{k-1, N-k}$
Total	S_T^2	$N-1$		

Following one-way classified ANOVA, we get

$$\text{Total Sum of Square (TSS)} = S_T^2 = \sum_{i=1}^{k} \sum_{j=1}^{n_i} \left(X_{ij} - \overline{X}_{..} \right)^2,$$

$$\text{Sum of Squares of Errors (SSE)} = S_E^2 = \sum_{i=1}^{k} \sum_{j=1}^{n_i} \left(X_{ij} - \overline{X}_{i.} \right)^2,$$

$$\text{Sum of Squares due to Treatment (SST)} = S_t^2 = \sum_{i=1}^{k} n_i \left(\overline{X}_{i.} - \overline{X}_{..} \right)^2,$$

$$\text{Mean sum of squares due to treatments} = \text{MST} = \frac{S_t^2}{k-1},$$

$$\text{Mean sum of squares due to error} = \text{MSE} = \frac{S_E^2}{N-k}.$$

We get a CRD table as shown in the Table 10.3.2.

■ Example 1

Gene Therapy (Human Genome Project Information)

Gene therapy is a technique for correcting defective genes responsible for disease development. Researchers may use one of several approaches for correcting faulty genes:

Technique A: A normal gene may be inserted into a nonspecific location within the genome to replace a nonfunctional gene. This approach is the most common.

Technique B: An abnormal gene could be swapped for a normal gene through homologous recombination.

Technique C: The abnormal gene could be repaired through selective reverse mutation, which returns the gene to its normal function.

Technique D: The regulation (the degree to which a gene is turned on or off) of a particular gene could be altered.

Table 10.3.3 Gene Therapy

	Subjects				
	1	2	3	4	5
Treatment A	96	99	100	104	84
Treatment B	93	90	75	80	90
Treatment C	70	90	84	76	78
Treatment D	78	87	67	66	76

Table 10.3.3 shows the data set obtained from the gene expression levels in a target cell by using these four different treatments to subjects. Check whether there is a statistically significant difference between various treatments.

Solution

The null hypothesis in this case is that all the treatments are the same. We obtain the following table using the given data:

SUMMARY				
Groups	Count	Sum	Average	Variance
Treatment A	5	483	96.6	57.8
Treatment B	5	428	85.6	59.3
Treatment C	5	398	79.6	58.8
Treatment D	5	374	74.8	74.7

ANOVA						
Source of Variation	SS	df	MS	F	P-value	F critical
Between Groups (Between Different Treatments)	1326.15	3	442.05	7.055866	0.003092	3.238871522
Within Groups (Errors)	1002.4	16	62.65			
Total	2328.55	19				

Since the p-value is $0.003092 < \alpha = 0.05$, we reject the claim that the treatments are the same. Therefore, on the basis of the given sample, we conclude that all the treatments are statistically significant from each other and hence produce different expression levels in a targeted cell.

The following R-code performs CRD for the data given in example 1:

```
> #Example 1
> rm(list = ls( ))
>
> genetherapy <- data.frame(expr = c(96, 99, 100, 104, 84,
+ 93, 90, 75, 80, 90,
+ 70, 90, 84, 76, 78,
+ 78, 87, 67, 66, 76),
+ subjects = factor(c(rep("TreatmentA",5), rep("TreatmentB", 5),
+ rep("TreatmentC", 5), rep("TreatmentD", 5))))
> genetherapy
expr subjects
1 96 TreatmentA
2 99 TreatmentA
3 100 TreatmentA
4 104 TreatmentA
5 84 TreatmentA
6 93 TreatmentB
7 90 TreatmentB
8 75 TreatmentB
9 80 TreatmentB
10 90 TreatmentB
11 70 TreatmentC
12 90 TreatmentC
13 84 TreatmentC
14 76 TreatmentC
15 78 TreatmentC
16 78 TreatmentD
17 87 TreatmentD
18 67 TreatmentD
19 66 TreatmentD
20 76 TreatmentD
>
> # mean, variance and standard deviation
>
> sapply(split(genetherapy$expr,genetherapy$subjects),mean)
TreatmentA TreatmentB TreatmentC TreatmentD
96.6       85.6       79.6       74.8
>
> sapply(split(genetherapy$expr,genetherapy$subjects),var)
TreatmentA TreatmentB TreatmentC TreatmentD
57.8       59.3       58.8       74.7
```

```
>
> sapply(split(genetherapy$expr,genetherapy$subjects),sd)
TreatmentA TreatmentB TreatmentC TreatmentD
7.602631    7.700649    7.668116    8.642916
>
> #Box-Plot
> # See Figure 10.3.1
> boxplot(split(genetherapy$expr,genetherapy$subjects),xlab = "Subjects",ylab
= "Expression Levels",col = "red")
>
> # lm( ) to fit a Linear Model
>
> fitgene<- lm(expr~subjects, data = genetherapy)
>
> #Displays the result
>
> summary(fitgene)
Call:
lm(formula = expr ~ subjects, data = genetherapy)
Residuals:
Min    1Q    Median  3Q    Max
-12.60 -6.15 1.80      4.40  12.20
Coefficients:
Estimate Std. Error t value Pr(>|t|)
(Intercept)    96.600    3.540 27.290 7.59e-15 ***
subjectsTreatmentB -11.000    5.006 -2.197 0.043066 *
subjectsTreatmentC -17.000    5.006 -3.396 0.003692 **
subjectsTreatmentD -21.800    5.006 -4.355 0.000491 ***
—
Signif. codes: 0 '***' 0.001 '**' 0.01 '*' 0.05 '.' 0.1 ' ' 1
Residual standard error: 7.915 on 16 degrees of freedom
Multiple R-squared: 0.5695, Adjusted R-squared: 0.4888
F-statistic: 7.056 on 3 and 16 DF, p-value: 0.003092
> #Use anova( )
>
> anova(fitgene)
Analysis of Variance Table
Response: expr
Df    Sum Sq Mean Sq F value Pr(>F)
subjects 3    1326.15 442.05    7.0559    0.003092 **
Residuals 16    1002.40 62.65
—
Signif. codes: 0 '***' 0.001 '**' 0.01 '*' 0.05 '.' 0.1 ' ' 1
>
```

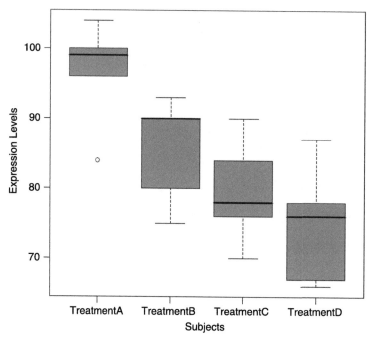

FIGURE 10.3.1

Box plot for subjects and expression levels.

10.4 RANDOMIZED BLOCK DESIGN

In the completely randomized design, we found that the CRD becomes less informative if the experimental material is not homogenous. In such situations, blocking is used to divide the whole experimental material into homogeneous strata or subgroups, which are known as blocks. After blocking, if the treatments are applied to experimental units randomly, this design is called a randomized block design (RBD). The randomization is restricted within the blocks in the RBD. Each treatment has the same number of replications. We assume that there is no interaction between treatments and blocks. The RBD is a good method to analyze the data when the experiments are spread over a period of time or space and the possibility of systematic variation or trends exists. The time series data obtained using the microarray analysis are spread over time, and therefore, the RBD may be suitable for further analysis. In the RBD, there is no restriction on the number of replicates needed or treatment needed, but in general, at least *two* replicates are required to carry out the test of significance. Also some other treatments may be included more than once without making the analysis a bit more complicated. The error of any treatment can be isolated, and any number of treatments may be omitted from the analysis without complicating the analysis.

Table 10.4.1 Two-Way Classified Data

		Classification-2 (Blocks)			Means	Total
		1	2 ...	h		
Classification-1	1	X_{11}	X_{12} ...	X_{1h}	$\overline{X}_{1.}$	$T_{1.}$
	2	X_{21}	X_{22} ...	X_{2h}	$\overline{X}_{2.}$	$T_{2.}$
(Treatments)		.	.	.	.	.
		.	.	.	.	.
		.	.	.	.	.
	i	X_{i1}	X_{i2} ...	X_{ih}	$\overline{X}_{i.}$	$T_{i.}$
		.	.	.	.	.
		.	.	.	.	.
		.	.	.	.	.
	k	X_{k1}	X_{k2} ...	X_{kh}	$\overline{X}_{k.}$	$T_{k.}$
Means		$\overline{X}_{.1}$	$\overline{X}_{.2}$ ...	$\overline{X}_{.h}$	$\overline{X}_{..}$	
						G
Total		$T_{.1}$	$T_{.2}$ ...	$T_{.h}$		

The RBD is not recommended when a large number of treatments and blocks have considerable variability.

The layout of the RBD is the same as that of the two-way classified ANOVA. Using the same notation as that of Chapter 9, Table 10.4.1 gives the classification of data for the RBD. In tabular form, we represent the two-way classification of data as follows.

For the RBD, we consider the linear model

$$X_{ij} = \mu + \tau_i + b_j + \varepsilon_{ij}, \quad i = 1, 2, \ldots, k; j = 1, 2, \ldots, h. \quad (10.4.1)$$

Here we define μ as the general mean effect given by

$$\mu = \frac{\sum_{i=1}^{k} \sum_{j=1}^{h} \mu_{ij}}{N},$$

where μ_{ij} is the fixed effect due to ith treatment (condition or effect or tissue type, etc.) and jth block.

α_i is the effect of the ith treatment given by

$$\alpha_i = \mu_{i.} - \mu, \quad \text{with} \quad \mu_{i.} = \frac{\sum_{j=1}^{h} \mu_{ij}}{h},$$

$$\beta_j = \mu_{.j} - \mu, \quad \text{with} \quad \mu_{.j} = \frac{\sum_{j=1}^{k} \mu_{ij}}{k},$$

$$\gamma_{ij} = \mu_{ij} - \mu_{i.} - \mu_{.j} + \mu,$$

and ε_{ij} is the random error effect.

We make the following assumptions about the model (10.4.1):

1. All the observations X_{ij} are independent.

2. Effects are additive in nature.

3. The random effects are independently and identically distributed as $N(0, \sigma_\varepsilon^2)$.

We would like to test the equality of the population means, that is, that mean effects of all the treatments as well as blocks are the same. We can state it as follows:

$$H_0 : \begin{cases} H_t : \tau_1 = \tau_2 = \ldots = \tau_k = 0, \\ H_b : b_1 = b_2 = \ldots = b_k = 0. \end{cases} \tag{10.4.2}$$

We follow terms used in Chapter 9 and let

$$\text{Total Sum of Square (TSS)} = S_T^2 = \sum_{i=1}^{k} \sum_{j=1}^{h} (X_{ij} - \overline{X}_{..})^2,$$

$$\text{Sum of Squares of Errors (SSE)} = S_E^2 = \sum_{i=1}^{k} \sum_{j=1}^{h} (X_{ij} - \overline{X}_{i.} - \overline{X}_{.j} - \overline{X}_{..})^2,$$

$$\text{Sum of Squares due to Treatment (SST)} = S_t^2 = h \sum_{i=1}^{k} (\overline{X}_{i.} - \overline{X}_{..})^2,$$

$$\text{Sum of Squares due to Blocks (SSB)} = S_b^2 = k \sum_{j=1}^{h} (\overline{X}_{.j} - \overline{X}_{..})^2.$$

To test the null hypothesis, H_t (10.4.2), we use the F-test given by

$$F = \frac{\text{MST}}{\text{MSE}} \sim F_{k-1,(k-1)(h-1)}.$$

Table 10.4.2 ANOVA Table for the RBD

Source of Variations	Sum of Squares	Degrees of Freedom	Mean Sum of Squares	Variance Ratio
Treatment	S_t^2	$k - 1$	$MST = \frac{S_t^2}{k-1}$	$F = \frac{MST}{MSE} \sim F_{k-1,(k-1)(h-1)}$
Blocks	S_b^2	$h - 1$	$MSB = \frac{S_b^2}{h-1}$	$F = \frac{MSB}{MSE} \sim F_{h-1,(k-1)(h-1)}$
Error	S_E^2	$(k-1)(h-1)$	$MSE = \frac{S_E^2}{(k-1)(h-1)}$	
Total	S_T^2	$N - 1$		

and to test the null hypothesis, H_b (10.4.2), we use the F-test given by

$$F = \frac{MSB}{MSE} \sim F_{h-1,(k-1)(h-1)},$$

where

Mean sum of squares due to treatments $= MST = \frac{S_t^2}{k-1}$,

Mean sum of squares due to blocks (varieties) $= MSB = \frac{S_b^2}{h-1}$,

Mean sum of squares due to errors $= MSE = \frac{S_E^2}{(k-1)(h-1)}$.

If the null hypothesis is true, the computed value of F must be close to 1; otherwise, it will be greater than 1, which will lead to rejection of the null hypothesis.

We can now form the ANOVA table for the RBD as shown in Table 10.4.2.

■ Example 1

Drosophila melanogaster BX-C serves as a model system for studying complex gene regulation. It is known that a cis-regulatory region of nearly 300 kb controls the expression of the three bithorax complex (BX-C) homeotic genes: Ubx, abd-A, and Abd-B1, 2. Table 10.4.3 shows the effects of individual factors in terms of the BX-C position and the tissue type in terms of mean relative methylation levels above background (RMoB). Flies were grown at 22.1° Centigrade. Five flies of each genotype were collected, and DNA was extracted separately from each six sets of three adult heads or three adult abdomens.

Using the RBD, test the hypothesis that there is no significant difference between BX-C positions and the tissue types.

Table 10.4.3 BX-C Position and Tissue Type

Tissue Type (b_i)	BX-C position (τ_i)					
	1	2	3	4	5	6
Abdomen	0.21	0.35	0.65	0.97	1.25	1.01
Head	0.15	0.20	0.75	1.10	0.9	0.95

Solution

We would like to test the hypothesis that the mean effects of all the positions as well as tissue type are the same. We can state this hypothesis as follows:

$$H_0 : \begin{cases} H_t : \tau_1 = \tau_2 = \ldots = \tau_6 = 0, \\ H_b : b_1 = b_2 = 0. \end{cases}$$

We then obtain the following table:

ANOVA: RBD				
SUMMARY	**Count**	**Sum**	**Average**	**Variance**
Abdomen	6	4.44	0.74	0.1654
Head	6	4.05	0.675	0.16275
BX-C-1	2	0.36	0.18	0.0018
BX-C-2	2	0.55	0.275	0.01125
BX-C-3	2	1.4	0.7	0.005
BX-C-4	2	2.07	1.035	0.00845
BX-C-5	2	2.15	1.075	0.06125
BX-C-6	2	1.96	0.98	0.0018

ANOVA						
Source of Variation	**SS**	**df**	**MS**	**F**	**P-value**	**F critical**
Tissue Type	0.012675	1	0.012675	0.82439	0.405534	6.607891
BX-C	1.563875	5	0.312775	20.34309	0.002453	5.050329
Error	0.076875	5	0.015375			
Total	1.653425	11				

Since the p-value for the tissue type is $0.405534 > \alpha = 0.05$, we fail to reject the claim that the mean effect of tissue type on the mean relative methylation levels above background (RMoB) is zero.

We notice that the mean effect of the BX-C position on the mean relative methylation levels above background (RMoB) is statistically significant because the p-value $= 0.002453 < \alpha = 0.05$.

The following R-code provides the RBD analysis for example 1:

```
> #Example 1
> rm(list = ls( ))
> # Tissue type "Abdomen" is represented by 1
> # Tissue type "Head" is represented by 2
>
> effect <- data.frame(mlevel = c(0.21, 0.35, 0.65, 0.97, 1.25, 1.01,
+ 0.15, 0.20, 0.75, 1.10, 0.9, 0.95),
+ tissuetype = factor(rep(rep(1:2, rep(6,2)),1))
+ ,BXCposition = factor(rep(c("1","2","3","4","5","6"), c(1,1,1,1,1,1))))
> effect
mlevel tissuetype BXCposition
1    0.21    1    1
2    0.35    1    2
3    0.65    1    3
4    0.97    1    4
5    1.25    1    5
6    1.01    1    6
7    0.15    2    1
8    0.20    2    2
9    0.75    2    3
10   1.10    2    4
11   0.90    2    5
12   0.95    2    6
>
>
> # mean, variance and standard deviation
>
> sapply(split(effect$mlevel,effect$tissuetype), mean)
1     2
0.740 0.675
>
> sapply(split(effect$mlevel,effect$BXCposition), mean)
1     2     3     4     5     6
0.180   0.275   0.700   1.035   1.075   0.980
>
> sapply(split(effect$mlevel,effect$tissuetype), var)
1     2
0.16540   0.16275
>
```

```
> sapply(split(effect$mlevel,effect$BXCposition), var)
1        2         3         4         5        6
0.00180   0.01125   0.00500   0.00845   0.06125   0.00180
>
> sapply(split(effect$mlevel,effect$tissuetype), sd)
1       2
0.4066940 0.4034229
>
> sapply(split(effect$mlevel,effect$BXCposition), sd)
1        2         3         4         5
0.04242641   0.10606602   0.07071068   0.09192388   0.24748737
6
0.04242641
>
> #Box-Plot
>
> boxplot(split(effect$mlevel,effect$tissuetype),xlab = "Tissue type",ylab = "Methyla-
tion Level",col = "blue")
> boxplot(split(effect$mlevel,effect$BXCposition),xlab = "BX-C position",ylab =
"Methylation level",col = "red")
> # See Figure 10.4.1, and Figure 10.4.2
> fitexp<-lm(mlevel~tissuetype + BXCposition, data = effect)
> #Display Results
> fitexp
Call:
lm(formula = mlevel ~ tissuetype + BXCposition, data = effect)
Coefficients:
(Intercept) tissuetype2 BXCposition2 BXCposition3 BXCposition4
0.2125    -0.0650   0.0950    0.5200    0.8550
BXCposition5 BXCposition6
0.8950    0.8000
> #Use anova( )
>
> anova(fitexp)
Analysis of Variance Table
Response: mlevel
```

	Df	Sum Sq	Mean Sq	F value	Pr(>F)
tissuetype	1	0.01268	0.01268	0.8244	0.405534
BXCposition	5	1.56387	0.31277	20.3431	0.002453 **
Residuals	5	0.07688	0.01538		

—

Signif. codes: 0 '***' 0.001 '**' 0.01 '*' 0.05 '.' 0.1 ' ' 1   ■

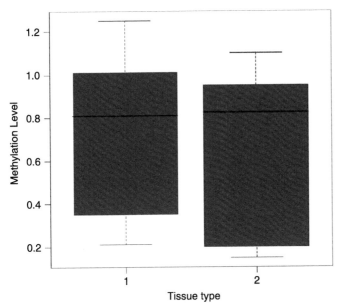

FIGURE 10.4.1

Box plot for tissue type and methylation level.

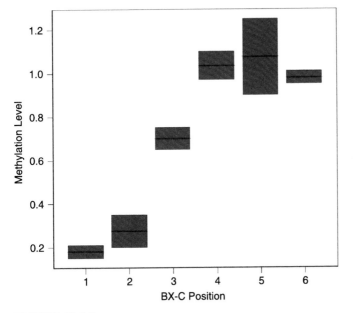

FIGURE 10.4.2

Box plot for BX-C position and methylation level.

10.5 LATIN SQUARE DESIGN

In the randomized block design, the experimental material is divided into relatively homogeneous blocks, and treatments are applied randomly within the blocks. However, sometimes the experimental material has inherent variability parallel to blocks. This will make the RBD inefficient because the design will not be able to control the variations. Practically, it is difficult to know the direction of the variability in the experimental material.

To overcome this problem, we use a design known as the Latin square design (LSD), which controls variation in two perpendicular directions. In the LSD, the number of treatments is equal to the number of replications. If we want to apply m treatments, we must have $m \times m = m^2$ experimental units. The experimental material is divided into m^2 experimental units arranged in a square so that each row as well as each column contains m units. Next, we allocate m treatments at random to these rows and columns in such a way that every treatment occurs once and only once in each row and in each column. For example, say we have four treatments on which we would like to study the effect of two factors. Then the LSD for such a case is given in Table 10.5.1.

The LSD makes an assumption that there is no interaction between factors, which may not be true in general. The LSD requires that the number of treatments is equal to the number of replications that limit its application. The LSD is suitable when we have a number of treatments between 5 and 10, but if the number of treatments is more than 10 or 12, the design becomes too large and does not remain homogeneous. Kerr et al. (2000) used a data set to show the working of the LSD in the microarray experiment settings. Kerr et al. used two arrays such that on array 1 the liver sample is assigned to the "red" dye, and the muscle sample is assigned to the "green" dye. On array 2, the dye assignments were reversed.

Dye	Array 1	Array 2
Red	Liver	Muscle
Green	Muscle	Liver

Let Y_{ijk} $(i, j, k = 1, 2, \ldots, m)$ be the measurement obtained from the experimental unit in the ith row, jth column, and receiving the kth treatment. Thus, there are only m^2 of the possible m^2 values of experimental units, which can be allocated to the design. We formulate the linear additive model, for an observation made per experimental unit, as

$$y_{ijk} = \mu + \alpha_i + \beta_j + \tau_k + \varepsilon_{ijk} \tag{10.5.1}$$

Table 10.5.1 4 × 4 Latin Square Design

	Factor 1			
	A	C	B	D
Factor 1	B	D	A	C
	D	A	C	B
	C	B	D	A

where μ is the overall mean effect; α_i, β_j, and τ_k are the effects due to the ith row (ith level of factor 1), jth column (jth level of factor 2) and kth treatment, respectively; and ε_{ijk} is random error assumed to be distributed as $\varepsilon_{ijk} \sim N(0, \sigma_e^2)$.

The null hypotheses that we want to test are

$$H_0 : \begin{cases} H_\alpha : \alpha_1 = \alpha_2 = \ldots = \alpha_m = 0, \\ H_\beta : \beta_1 = \beta_2 = \ldots \beta_m = 0, \\ H_\tau : \tau_1 = \tau_2 = \ldots \tau_m = 0. \end{cases} \qquad (10.5.2)$$

We use the following notations:

$G = y\ldots = $ Grand total of all the m^2 observations,

$\bar{y}_{\ldots} = $ Mean of all the m^2 observations,

$R_i = y_{i..} = $ Total of m observations in the ith row,

$C_j = y_{.j.} = $ Total of m observations in the jth column,

$T_k = y_{..k} = $ Total of m observations in the kth treatment.

Then we have

$$\sum_{i=1}^{m}\sum_{j=1}^{m}\sum_{k=1}^{m}(y_{ijk} - \bar{y}_{\ldots})^2 = \sum_{i=1}^{m}\sum_{j=1}^{m}\sum_{k=1}^{m}\{(\bar{y}_{i..} - \bar{y}_{\ldots}) + (\bar{y}_{.j.} - \bar{y}_{\ldots}) + (\bar{y}_{..k} - \bar{y}_{\ldots})$$
$$+ (y_{ijk} - \bar{y}_{i..} - \bar{y}_{.j.} - \bar{y}_{..k} + 2\bar{y}_{\ldots})\}^2$$
$$= m\sum_{i=1}^{m}(\bar{y}_{i..} - \bar{y}_{\ldots})^2 + m\sum_{j=1}^{m}(\bar{y}_{.j.} - \bar{y}_{\ldots})^2 + m\sum_{k=1}^{m}(\bar{y}_{..k} - \bar{y}_{\ldots})^2$$
$$+ \sum_{i=1}^{m}\sum_{j=1}^{m}\sum_{k=1}^{m}(y_{ijk} - \bar{y}_{i..} - \bar{y}_{.j.} - \bar{y}_{..k} + 2\bar{y}_{\ldots})^2.$$
$$(10.5.3)$$

Since the product terms are a sum of deviations from their means, the product terms are equal to zero. The equation (10.5.3) is written as

$$\text{TSS} = \text{SSR} + \text{SSC} + \text{SST} + \text{SSE}$$

where

$$\text{TSS} = S_T^2 = \text{Total sum of squares} = \sum_{i=1}^{m}\sum_{j=1}^{m}\sum_{k=1}^{m}(y_{ijk} - \bar{y}_{...})^2,$$

$$\text{SSR} = S_R^2 = \text{Sum of squares due to rows} = m\sum_{i=1}^{m}(\bar{y}_{i..} - \bar{y}_{...})^2,$$

$$\text{SSC} = S_C^2 = \text{Sum of squares due to columns} = m\sum_{j=1}^{m}(\bar{y}_{.j.} - \bar{y}_{...})^2,$$

$$\text{SST} = S_t^2 = \text{Sum of squares due to treatments} = m\sum_{k=1}^{m}(\bar{y}_{..k} - \bar{y}_{...})^2,$$

$$\text{SSE} = S_E^2 = \text{TSS} - \text{SSR} - \text{SSC} - \text{SST}.$$

The least square estimates of parameters μ, α_i, β_j, and τ_k, $i = 1, 2, \ldots, m$ in (10.5.2) are obtained by minimizing the residual sum of squares E given by

$$E = \sum_{i=1}^{m}\sum_{j=1}^{m}\sum_{k=1}^{m}(y_{ijk} - \mu - \alpha_i - \beta_i - \tau_k)^2. \qquad (10.5.4)$$

When we use the principle of least squares, the normal equations for estimating μ, α_i, β_j and τ_k are given by

$$\frac{\partial E}{\partial \mu} = 0, \ \frac{\partial E}{\partial \alpha_i} = 0, \ \frac{\partial E}{\partial \beta_j} = 0, \ \frac{\partial E}{\partial \tau_k} = 0. \qquad (10.5.5)$$

If $\hat{\mu}$, $\hat{\alpha}_i$, $\hat{\beta}_j$ and $\hat{\tau}_k$ are the estimates of μ, α_i, β_j, and τ_k and if we assume that

$$\sum_i \hat{\alpha}_i = \sum_j \hat{\beta}_j = \sum_k \hat{\tau}_k = 0 \qquad (10.5.6)$$

then

$$\left.\begin{array}{l} y_{...} = m^2\hat{\mu} \qquad \Rightarrow \hat{\mu} = \bar{y}_{...}, \\ y_{i..} = m(\hat{\mu} + \hat{\alpha}_i) \Rightarrow \hat{\alpha}_i = \bar{y}_{i..} - \bar{y}_{...}, \\ y_{.j.} = m(\hat{\mu} + \hat{\beta}_j) \Rightarrow \hat{\beta}_j = \bar{y}_{.j.} - \bar{y}_{...}, \\ y_{..k} = m(\hat{\mu} + \hat{\tau}_k) \Rightarrow \hat{\tau}_k = \bar{y}_{..k} - \bar{y}_{...} \end{array}\right\} \qquad (10.5.7)$$

Then the ANOVA table is created for the LSD as shown in Table 10.5.2.

Table 10.5.2 ANOVA for the Latin Square Design

Source of Variations	Degree of Freedom	Sum of Squares	Mean Sum of Squares	Variance Ratio "F"
Rows	$m - 1$	S_R^2	$MSR = \frac{S_R^2}{m-1}$	$F_R = \frac{MSR}{MSE}$
Columns	$m - 1$	S_C^2	$MSC = \frac{S_C^2}{m-1}$	$F_C = \frac{MSC}{MSE}$
Treatments	$m - 1$	S_t^2	$MST = \frac{S_t^2}{m-1}$	$F_t = \frac{MST}{MSE}$
Error	$(m - 1).(m - 2)$	S_E^2	$MSE = \frac{S_E^2}{(m-1)(m-2)}$	
Total	$m^2 - 1$			

■ **Example 1**

During the testing and training of explosives at military bases, some explosives remain unexploded in the field and are not collected back due to security concerns. This unexploded material mixes in the environment through rain and other natural processes. The long-term effect of the unexploded material on gene expressions of species is of great importance. An experiment was carried out to determine the effect of leftover RDX on the gene expression levels of the zebra fish.

Four different doses of RDX:

A: Zero mg of RDX, B: 1 mg of RDX, C: 2 mg of RDX, and D: 3 mg of RDX.

Duration of exposure to RDX:

D-1: 1 minute, D-2: 2 minutes, D-3: 4 minutes, and D-4: 6 minutes.

Sample tissue type:

Type-1: Head, Type-2: muscle, Type-3: Fin, and Type-4: tail.

The gene expression measurements are given in Table 10.5.3. Use the LSD to find whether there is significant effect of RDX on the gene expression level.

Solution

We set up the null hypotheses as follows:

$$H_0 : \begin{cases} H_\alpha : \alpha_1 = \alpha_2 = \ldots \alpha_m = 0, \\ H_\beta : \beta_1 = \beta_2 = \ldots \beta_m = 0, \\ H_\tau : \tau_1 = \tau_2 = \ldots = \tau_m = 0, \end{cases} \qquad (10.5.8)$$

where α_i, β_j, and τ_k are the effects due to ith tissue, jth duration of time, and kth treatment (dosage of RDX), respectively.

Table 10.5.3 Effect of RDX on the Gene Expression Measurements

| Tissue type | Time of Exposure | | | |
	Time-1	Time-2	Time-3	Time-4
Type-1	D	B	C	A
	90	94	86	78
Type-2	C	A	D	B
	82	76	94	90
Type-3	A	D	B	C
	65	98	95	81
Type-4	B	C	A	D
	91	85	71	99

Table 10.5.4 ANOVA for the LSD

Source of Variation	Degrees of Freedom	Sum of Squares	Mean Sum of Squares	Variance Ratio	P-value
Tissue	3	12.188	4.062	0.289	0.832
Time	3	89.187	29.729	2.114	0.200
RDX	3	1265.187	421.729	29.990	0.001
Error	6	84.375	14.062		
Total	15	1450.938			

Table 10.5.4 is obtained for the ANOVA using the LSD.

Thus, we find that the effects of tissue and time are not significant, but the treatment is statistically significant. Thus, different dosages of RDX have a significant effect on the gene expression levels of the zebra fish.

Following is the R-code for the LSD analysis required in example 1:

```
> #Example 1
> rm(list = ls( ))
> # Tissue type "Type-1" is represented by 1, "Type-2" by 2, and so on
> # Time of Exposure "Time-1" is represented by 1, "Time-2" is represented by 2, and
so on
> # Dose level is represented by "A", "B", "C", "D"
>
> effect <- data.frame(exp = c(90, 94, 86, 78,
+ 82, 76, 94, 90,
+ 65, 98, 95, 81,
+ 91, 85, 71, 99),
```

```
+ tissuetype = factor(rep(rep(1:4, rep(4,4)),1))
+ ,time = factor(rep(c("Time-1","Time-2","Time-3","Time-4"), c(1,1,1,1)))
+ ,
+ dose = factor(rep(c("D","B","C","A",
+ "C", "A", "D", "B",
+ "A", "D", "B", "C",
+ "B", "C", "A", "D"), c(1,1,1,1, 1,1,1,1, 1,1,1,1, 1,1,1,1))))
>
>
> effect
exp   tissuetype    time dose
1    90  1 Time-1   D
2    94  1 Time-2   B
3    86  1 Time-3   C
4    78  1 Time-4   A
5    82  2 Time-1   C
6    76  2 Time-2   A
7    94  2 Time-3   D
8    90  2 Time-4   B
9    65  3 Time-1   A
10   98  3 Time-2   D
11   95  3 Time-3   B
12   81  3 Time-4   C
13   91  4 Time-1   B
14   85  4 Time-2   C
15   71  4 Time-3   A
16   99  4 Time-4   D
>
>
> # mean, and variance
>
> sapply(split(effect$exp,effect$tissuetype), mean)
1     2     3     4
87.00 85.50 84.75 86.50
>
> sapply(split(effect$exp,effect$time), mean)
Time-1 Time-2 Time-3 Time-4
82.00 88.25 86.50 87.00
>
> sapply(split(effect$exp,effect$dose), mean)
A     B     C     D
72.50 92.50 83.50 95.25
>
> sapply(split(effect$exp,effect$tissuetype), var)
1     2     3     4
```

```
46.66667 65.00000 228.25000 139.66667
>
> sapply(split(effect$exp,effect$time), var)
Time-1   Time-2   Time-3   Time-4
144.6667 96.2500 123.0000 90.0000
>
> sapply(split(effect$exp,effect$dose), var)
A        B         C         D
33.666667 5.666667 5.666667 16.916667
>
>
> #Box-Plot
>
> boxplot(split(effect$exp,effect$tissuetype),xlab = "Tissue type",ylab = "Expression
Level",col = "blue")
> boxplot(split(effect$exp,effect$time),xlab = "Time",ylab = "Expression Level",col =
"red")
> boxplot(split(effect$exp,effect$dose),xlab = "Dose",ylab = "Expression Level",col =
"green")
> # See Figure 10.5.1, Figure 10.5.2, and Figure 10.5.3
> fitexp<-lm(exp~tissuetype + time + dose, data = effect)
> #Display Results
> fitexp
Call:
lm(formula = exp ~ tissuetype + time + dose, data = effect)
Coefficients:
(Intercept) tissuetype2 tissuetype3 tissuetype4 timeTime-2 timeTime-3
69.62       -1.50       -2.25       -0.50       6.25       4.50
timeTime-4  doseB  doseC  doseD
5.00        20.00  11.00  22.75
> #Use anova( )
> anova(fitexp)
Analysis of Variance Table
Response: exp
Df Sum Sq Mean Sq F value Pr(>F)
tissuetype   3   12.19    4.06    0.2889 0.8322001
time     3   89.19   29.73    2.1141 0.1998049
dose     3 1265.19    421.73 29.9896 0.0005219 ***
Residuals    6   84.37    14.06

—
Signif. codes: 0 '***' 0.001 '**' 0.01 '*' 0.05 '.' 0.1 ' ' 1
>
```

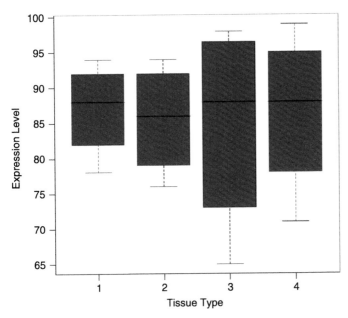

FIGURE 10.5.1

Box plot for tissue type and expression level.

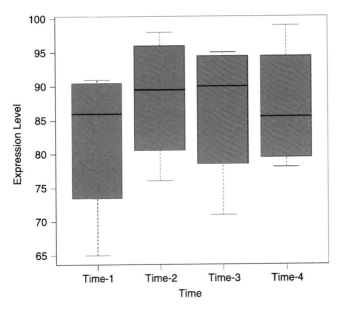

FIGURE 10.5.2

Box plot for time and expression level.

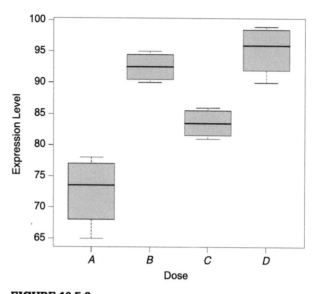

FIGURE 10.5.3

Box plot for dose level and expression level.

10.6 FACTORIAL EXPERIMENTS

The designs like CRD, RBD, and LSD were concerned with the comparison of the effects of a single set of treatments, for example, the effect of a medicine. However, if we would like to study effects of several factors of variation, these designs are not suitable. In *factorial experiments,* effects of several factors of variation are studied simultaneously. In factorial experiments, the treatments are all the combinations of different factors under study. In factorial experiments, the effects of each of the factors and the interaction effects, which are the variations in the effect of one factor as a result of different levels of other factors, are studied.

Suppose we want to study the effect of two types of hypothermia, say *A* and *B*, on cellular functions. Let there be *p* different types of hypothermia *A* (depending on the temperature) and there be *q* different types of hypothermia *B* (depending on the hours of exposure). Both types of hypothermia, *A* and *B*, affect human cellular function and produce broad changes in mRNA expression *in vitro*. The *p* and *q* are termed as the effectiveness of various treatments, namely, the different levels of hypothermia *A* and hypothermia *B*. If we conduct two simple experiments, one for hypothermia *A* and another for hypothermia *B*, they may be lengthy and costly and may not be useful because of unaccounted systematic changes in the background conditions. These single-factor experiments do not provide information regarding the relationship between two factors. Factorial experiments study the effects of several factors simultaneously. If n denotes the number of factors and s denotes the number of levels of each factor, we denote the factorial experiment as an s^n-factorial experiment. Thus, a 2^3-factorial experiment would imply that we have an experiment with three factors at two levels each, and a 3^2-factorial experiment would imply that we have an experiment with two factors at three levels each.

10.6.1 2^n-Factorial Experiment

In the 2^n-factorial experiment, there are n factors, each at two levels. The levels may be two quantitative levels or concentrations of factors. For example, if dose is a factor, then different levels of doses may be the levels of the factors. We may also consider the presence or absence of some characteristic as the two levels of a factor.

We consider two factors each at two levels $(0, 1)$. Thus, we have $2 \times 2 = 4$ treatment combinations in total. We use the following notation, given by Yates (1935). Let A and B denote two factors under study and 'a' and 'b' denote two different levels of the corresponding factors.

Four treatment combinations in a 2^2-factorial experiment can be listed as follows:

$a_1 b_1$ or '1': Factor A at the first level and factor B at the first level.

$a_2 b_1$ or 'a': Factor A at the second level and factor B at the first level.

$a_1 b_2$ or 'b': Factor A at the first level and factor B at the second level.

$a_2 b_2$ or 'ab': Factor A at the second level and factor B at the second level.

The ANOVA can be carried out by using either an RBD with r-replicates, with each replicate containing four units, or by using a 4×4 LSD. In the factorial experiments, there are separate tests for the interaction AB and separate tests for the main effects A and B. The sum of squares due to treatments is split into three orthogonal components A, B, and AB, each with one degree of freedom. The sum of squares due to treatment is equal to the sum of squares due to orthogonal components A, B, and AB.

Let there be r-replicates for the 2^2-factorial experiment and let $[1]$, $[a]$, $[b]$, and $[ab]$ denote the measurements obtained from r-replicates receiving treatments $1, a, b$, and ab, respectively. Let the corresponding mean values for r-replicates be (1), (a), (b), and (ab).

Using the orthogonal property, we obtain the factorial effect totals as

$$\left.\begin{array}{l} [A] = [ab] - [b] + [a] - [1] \\ [B] = [ab] + [b] - [a] - [1] \\ [AB] = [ab] - [a] - [b] + [1] \end{array}\right\}. \qquad (10.6.1)$$

The sum of squares due to any factorial effect are given by

$$\text{Sum of squares due to main effect } A = S_A^2 = \frac{[A]^2}{4r},$$

$$\text{Sum of squares due to main effect } B = S_B^2 = \frac{[B]^2}{4r},$$

$$\text{Sum of squares due to interaction } AB = S_{AB}^2 = \frac{[AB]^2}{4r}.$$

Here r is the number of replicates. The degree of freedom for each factorial effect is 1. As mentioned earlier, factorial experiments are carried out either as CRD, RBD, or LSD, except the fact that the treatment sum of squares is split into three orthogonal components each with one degree of freedom. We use RBD and create the classification of the data shown in Table 10.6.1.

We find

$$\text{Total Sum of Square (TSS)} = S_T^2 = \sum_{i=1}^{n} \sum_{j=1}^{n} \left(X_{ij} - \overline{X}_{..} \right)^2,$$

$$\text{Sum of Squares of Errors (SSE)} = S_E^2 = \sum_{i=1}^{n} \sum_{j=1}^{n} \left(X_{ij} - \overline{X}_{i.} - \overline{X}_{.j} - \overline{X}_{..} \right)^2,$$

$$\text{Sum of Squares due to Treatment (SST)} = S_t^2 = n \sum_{i=1}^{n} \left(\overline{X}_{i.} - \overline{X}_{..} \right)^2,$$

Table 10.6.1 n^2-Factorial with Two-Way Classification of the Data (RBD)

		Factor-2 (Treatments)				Means	Total
		1	2	$\cdots$	n		
Factor-1 (Blocks)	1	X_{11}	X_{12}	$\cdots$	X_{1n}	$\overline{X}_{1.}$	$T_{1.}$
	2	X_{21}	X_{22}	$\cdots$	X_{2n}	$\overline{X}_{2.}$	$T_{2.}$
	$\vdots$	$\vdots$	$\vdots$		$\vdots$	$\vdots$	$\vdots$
	i	X_{i1}	X_{i2}	$\cdots$	X_{in}	$\overline{X}_{i.}$	$T_{i.}$
	$\vdots$	$\vdots$	$\vdots$		$\vdots$	$\vdots$	$\vdots$
	n	X_{n1}	X_{n2}	$\cdots$	X_{nn}	$\overline{X}_{n.}$	$T_{n.}$
		$\overline{X}_{.1}$	$\overline{X}_{.2}$	$\cdots$	$\overline{X}_{.n}$	$\overline{X}_{..}$	
		$T_{.1}$	$T_{.2}$	$\cdots$	$T_{.n}$	$T_{..}$	G

Table 10.6.2 ANOVA for n^2-Factorial with RBD

Source of Variations	Sum of Squares	Degrees of Freedom	Mean Sum of Squares	Variance Ratio
Blocks	S_b^2	$n-1$	$MSB = \frac{S_b^2}{n-1}$	$F = \frac{MSB}{MSE} \sim F_{n-1,(n-1)(n-1)}$
Treatment	S_t^2	$n-1$	$MST = \frac{S_t^2}{n-1}$	$F = \frac{MST}{MSE} \sim F_{n-1,(n-1)(n-1)}$
$a_0 b_0$	$SS[1] = \frac{[1]^2}{4n}$	1	$MS[1] = SS[1]/1$	$F = \frac{MS[1]}{MSE} \sim F_{1,(n-1)(n-1)}$
$a_1 b_0$	$SS[A_1] = \frac{[A_1]^2}{4n}$	1	$MS[A_1] = SS[A_1]/1$	$F = \frac{MS[A_1]}{MSE} \sim F_{1,(n-1)(n-1)}$
$a_0 b_1$	$SS[B_1]/1 = \frac{[B_1]^2}{4n}$	1	$MS[B_1] = SS[B_1]/1$	$F = \frac{MS[B_1]}{MSE} \sim F_{1,(n-1)(n-1)}$
$\vdots$				
$a_n b_n$	$SS[A_n B_n] = \frac{[A_n B_n]^2}{4n}$	1	$MS[A_n B_n] = SS[A_n B_n]/1$	$F = \frac{MS[A_n B_n]}{MSE} \sim F_{1,(n-1)(n-1)}$
Error	S_E^2	$(n-1)(n-1)$	$MSB = \frac{S_E^2}{(n-1)(n-1)}$	
Total	S_T^2	$nn-1$		

Table 10.6.3 ANOVA for 2^n-Factorial with r Randomized Blocks

Source of Variations	Sum of Squares	Degrees of Freedom	Mean Sum of Squares	Variance Ratio
Blocks	S_b^2	$r-1$	$MSB = \frac{S_b^2}{r-1}$	$F = \frac{MSB}{MSE} \sim F_{r-1,(r-1)(2^n-1)}$
Treatment	S_t^2	2^n-1	$MST = \frac{S_t^2}{2^n-1}$	$F = \frac{MST}{MSE} \sim F_{r-1,(r-1)(2^n-1)}$
$a_0 b_0$	$SS[1] = \frac{[1]^2}{r2^n}$	1	$MS[1] = SS[1]/1$	$F = \frac{MS[1]}{MSE} \sim F_{1,(r-1)(2^n-1)}$
$a_1 b_0$	$SS[A_1] = \frac{[A_1]^2}{r2^n}$	1	$MS[A_1] = SS[A_1]/1$	$F = \frac{MS[A_1]}{MSE} \sim F_{1,(r-1)(2^n-1)}$
$a_0 b_1$	$SS[B_1] = \frac{[B_1]^2}{r2^n}$	1	$MS[B_1] = SS[B_1]/1$	$F = \frac{MS[B_1]}{MSE} \sim F_{1,(r-1)(2^n-1)}$
$\vdots$				
$a_1 b_1 c_1 \cdots k_1$	$SS[A_1 B_1 \cdots K_1] = \frac{[A_1 B_1 \cdots K_1]^2}{r2^n}$	1	$MS[A_1 B_1 \cdots K_1] = SS[A_1 B_1 \cdots K_1]/1$	$F = \frac{MS[A_1 B_1 \cdots K_1]}{MSE} \sim F_{1,(r-1)(2^n-1)}$
Error	S_E^2	$(r-1)(2^n-1)$	$MSB = \frac{S_E^2}{(r-1)(2^n-1)}$	
Total	S_T^2	$r2^n-1$		

$$\text{Sum of Squares due to blocks (SSB)} = S_b^2 = n\sum_{j=1}^{n}\left(\overline{X}_{.j} - \overline{X}_{..}\right)^2.$$

The ANOVA for the n^2-factorial using RBD is given in Table 10.6.2.

The ANOVA for 2^n-factorial with r randomized blocks is given in Table 10.6.3.

Here [1] is nothing but the grand total of effects, and therefore, it is not included in the ANOVA. The remaining main effects $a, b, c, d, \ldots,$ and interactions

$ab, abc, abcd, \ldots, abcd \ldots k$ are of much interest and therefore are included in the ANOVA for the factorial designs.

■ Example 1

In an experiment, the toxic effects of two chemicals, namely, A and B, are studied on their cellular functions. There are two levels of chemical A ($a_0 = 0$ mg of chemical A, $a_1 = 5$ mg of chemical A), and there are two levels of chemical B ($b_0 = 0$ mg of chemical B, $b_1 = 5$ mg of chemical B). All the combinations and levels of chemicals were studied in a randomized black design with four replications for each. Table 10.6.4 gives the gene expression levels for each combination. Determine whether there is a significant effect due to treatments.

Solution

The ANOVA for the 2^2-factorial design with RBD is given in Table 10.6.5.

From Table 10.6.5, we find that in all cases the p-value is less than $\alpha = 0.05$; therefore, we reject the null hypothesis that blocks as well as treatment differ

Table 10.6.4 Toxic Effects of Chemicals

Block	Treatment			
I	(1)	a	b	ab
	96	78	87	99
II	b	(1)	a	ab
	85	95	75	98
III	ab	b	(1)	a
	97	82	92	75
IV	a	ab	b	(1)
	77	97	85	92

Table 10.6.5 ANOVA for 2^2-Factorial Design with RBD on the Toxic Effects of Chemicals

Source of Variations	Sum of Squares	Degrees of Freedom	Mean Sum of Squares	Variance Ratio	P-value
Blocks	25.25	3	8.417	$F = 7.769$	0.007
Treatment	1106.75	3	368.917	$F = 340.538$	0.000
$a_1 b_0$	20.25	1	20.25	$F = 18.69$	0.0019
$a_0 b_1$	156.25	1	156.25	$F = 144.275$	0.0000
$a_1 b_1$	930.25	1	930.25	$F = 858.825$	0.0000
Error	9.75	9	1.083		
Total	1141.75	15			

Table 10.6.6 Toxic Effects of Chemicals in a Factorial Design					
A	**B**	**I**	**II**	**III**	**IV**
−1	−1	96	95	92	92
1	−1	78	75	75	77
−1	1	87	85	82	85
1	1	99	98	97	97

significantly. Thus, there is a significant effect due to treatments. Further, we see that the main effect due to combinations a_1b_0, a_0b_1, and interaction a_1b_1 have a significant effect on the gene expression levels.

To use the data in R, we need to rearrange as shown in Table 10.6.6.

R provides factorial analysis without incorporating RBD design, whereas the analysis reported in Table 10.6.4 is based on the analysis when RBD is incorporated in factorial design. The following analysis is obtained once we run the R-code for the factorial experiment in R, but note that these results are based on the fact that RBD is not incorporated in these codes.

```
> #Example 1
> rm(list = ls( ))
> # Enter Data in required format
>
> A <- c(-1, 1, -1,1)
> B <- c(-1, -1, 1, 1)
> I <- c(96, 95, 92, 92)
> II <- c(78, 75, 75, 77)
> III <- c(87, 85, 82, 85)
> IV <- c(99, 98, 97, 97)
>
> #Prepare Data Table
> data <- data.frame(A, B, I, II, III, IV )
> data
A B I II III IV
1 -1 -1 96 78 87 99
2 1 -1 95 75 85 98
3 -1 1 92 75 82 97
4 1 1 92 77 85 97
>
>
> # Compute sums for (1), (a), (b), (ab)
> sums <- apply(data[,3:6], 1, sum)
> names(sums) <- c("(1)", "(a)", "(b)", "(ab)")
> sums
(1) (a) (b) (ab)
360 353 346 351
```

```
> ymean <- sums/4
>
> # Make interaction plots
> par(mfrow = c(1,2))
> interaction.plot(A, B, ymean)
> interaction.plot(B, A, ymean)
>
> # See Figure 10.6.1.1 Prepare ANOVA table
>
> y <- c(I, II, III, IV)
> y
[1] 96 95 92 92 78 75 75 77 87 85 82 85 99 98 97 97
>
> factorA <- as.factor(rep(A,4))
> factorB <- as.factor(rep(B,4))
> factorA
[1] -1 1 -1 1 -1 1 -1 1 -1 1 -1 1 -1 1 -1 1
Levels: -1 1
> factorB
[1] -1 -1 1 1 -1 -1 1 1 -1 -1 1 1 -1 -1 1 1
Levels: -1 1
>
> model <- lm(y ~ factorA + factorB + factorA*factorB)
> anova(model)
Analysis of Variance Table
Response: y
Df  Sum Sq  Mean Sq  F value  Pr(>F)
factorA  1  0.25  0.25  0.0027  0.9595
factorB  1  16.00  16.00  0.1720  0.6857
factorA:factorB  1  9.00  9.00  0.0967  0.7611
Residuals  12  1116.50  93.04
>
>
> # Perform residual analysis
> windows( )
> par(mfrow = c(1,2))
> # Q-Q plot
> qqnorm(model$residuals)
> qqline(model$residuals)
>
> # residual plot
> plot(model$fitted.values,model$residuals)
> abline(h = 0)
> # See Figure 10.6.1.2
```

■

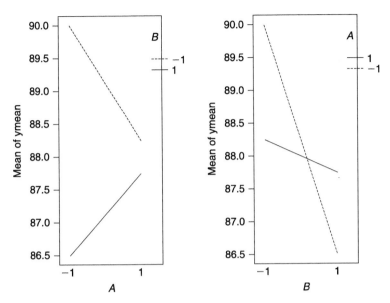

FIGURE 10.6.1.1

Interaction plot between: (a) A and B; (b) B and A.

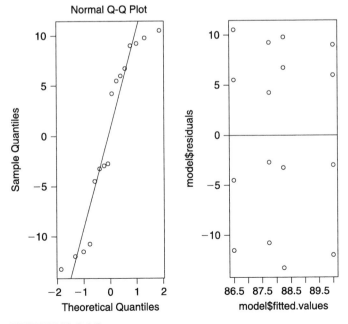

FIGURE 10.6.1.2

Normal Q-Q plot and residual plot.

10.7 REFERENCE DESIGNS AND LOOP DESIGNS

Experiments are becoming larger and larger, involving larger numbers of samples and conditions. It has increasingly becoming important to design experiments that are efficient as well as to provide more information about the parameters of the population. The design of experiments aims to minimize the inherent variations in the data and increase the precision of the parameters. If we wish to compare the number of conditions under a fixed sample size, we have two designs available: namely, reference design and loop design.

Reference design is the most commonly used design. Under this design, a reference sample is compared with each of the samples drawn from each of the conditions under study. In a microarray experiment, the reference array is hybridized along with the array under a condition. Thus, for each array under a condition, there will be a reference array. Therefore, the intensity of the reference array and condition array will get the same treatment all along the experiment and therefore will remove bias in the treatment. This relative hybridization intensity is already a standardized value and therefore can be used to compare other standardized values obtained from arrays under different conditions. Since one control or reference array is used per condition array, 50% of the hybridization resources are used to produce a control of reference arrays, which may not be useful in making inferences. Some of the reference designs are given in Figures 10.7.1 and 10.7.2. For the dye-swap microarray experiment, arrays are represented by an arrow; the tail of an arrow represents a green dye, and the head of the arrow represents a red dye.

	Treatment Samples		
Red Channel	R	R	R
Green Channel	T1	T2	T3

FIGURE 10.7.1

Classical reference design, with each treatment (condition) T1, T2, and T3, compared with a reference sample.

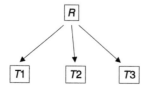

FIGURE 10.7.2

Classical reference design with one reference, R, (or control) sample and three treatment samples T1, T2, and T3.

Reference designs are robust in the sense that it is easy to add or remove any number of control samples. Reference designs are easy to plan and execute, and they are easy to analyze. Because of the simplicity of these designs, most biologists prefer reference designs. One reference design with dye swap and replications is shown in Figure 10.7.3.

Loop designs compare two or more conditions using a chain of conditions (see Figure 10.7.4). Loop designs remove the need for a reference sample for comparing various conditions, but one can use a reference sample in the loop anywhere. Since the loop design does not use control arrays in microarray experiments, only half of the arrays are required to attain the same sample size as the reference design. It is not easy to include new condition arrays or samples without destroying the original loop (see Figure 10.7.5).

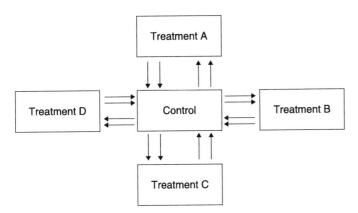

FIGURE 10.7.3

Reference design with dye swap and replications.

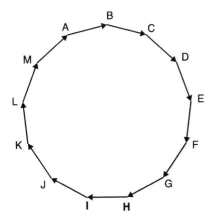

FIGURE 10.7.4

Loop design.

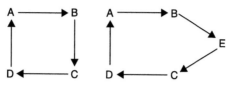

FIGURE 10.7.5

Adding a new treatment to a loop design.

FIGURE 10.7.6

A 2^2-factorial experiment incorporated into a loop design.

The factorial designs discussed in Section 10.6 could be incorporated into loop designs. For example, a 2^2-factorial design (two factors A and B; with two levels a and b) could be incorporated in the loop design as shown in Figure 10.7.6.

Interwoven loop designs are more complicated than simple loop designs. If we have five treatments with three biological replicates, it becomes complicated to design an interwoven loop design (see Figure 10.7.7). If we try to include time intervals for each treatment (Example 1 in section 10.5), it becomes a little more complicated, but the reference design still remains simple to design and analyze. Suppose we would like to compare XY treatment combinations, where $X = A, B, C, D,$ and E treatments and $Y = 1, 2, 3$ replicates. Each treatment combination needs to be compared with two other treatment combinations. The interwoven loop design is shown in Figure 10.7.7.

An interwoven loop with three reference samples $R1$, $R2$, and $R3$ and three treatments A, B, and C, with three replications each, is shown in Figure 10.7.8.

A more complicated loop design with replication and dye swap is shown in Figure 10.7.9.

The block design could also be incorporated in the loop design. For example, we would like to allocate eight treatments to two blocks, each block getting four treatments. This scenario is represented in Figure 10.7.10, where a solid line denotes one block and the dotted line denotes the other block. Treatments are represented by $A, B, \ldots, H$.

A block design with replications could be easily incorporated into the loop design. Suppose we have four blocks, denoted by different types of lines, and eight treatments, denoted by $A, B, \ldots, H$; then the loop design shown in Figure 10.7.10 could be modified into the loop design shown in Figure 10.7.11.

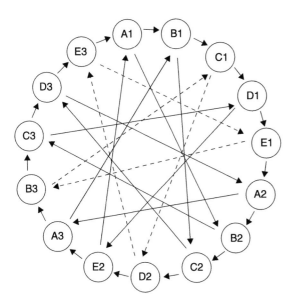

FIGURE 10.7.7

Interwoven loop design for five treatments with three independent biological replicates.

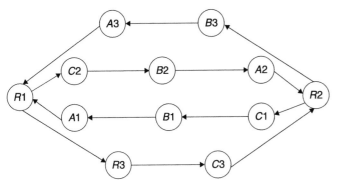

FIGURE 10.7.8

Interwoven design with replication and reference samples.

Each design has a specific application. We note that for class discovery, a reference design is preferable because of large gains in cluster performance. On the other hand, for the class comparisons, if we have a fixed number of arrays, a block design is more efficient than a loop or reference design, but the block design cannot be applied to clustering (Dobbin and Simon, 2002). If we have a fixed number of specimens, a reference design is more efficient than a loop or block design when intraclass variance is large. Most of the experiments in microarrays deal with dual-label problems. Two labels reduce the spot-to-spot

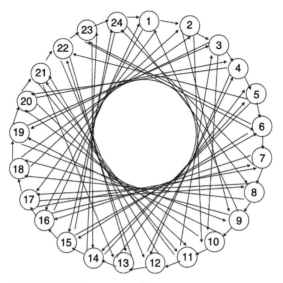

FIGURE 10.7.9

Loop design with dye swap and replication.

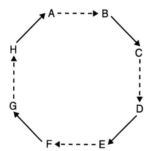

FIGURE 10.7.10

Complete block design with two blocks and eight treatments, incorporated in a loop design.

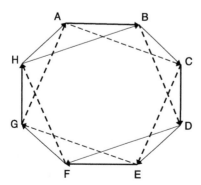

FIGURE 10.7.11

A block design with four blocks and eight treatments, incorporated in a loop design.

variations and provide more power in class-comparison experiments. The reference designs are labeled with the same dye. The spots on the reference design provide the estimate of "spot" effect for every spot on the array. Thus, we can correct spot-to-spot variation, which occurs in different arrays.

If, in a microarray experiment, arrays are limited in number, for the class comparison experiment, a balanced block design can be significantly more efficient than a reference design, and differentially expressed genes will have fewer false positive and false negative genes.

Let y_{ijkg} be the intensity from array i and dye j represent variety k and gene g. We can assume that same spot is spotted on each array in an experiment. Thus, gene effects are orthogonal to all effects of factors. We have two categories of effects: global effects ($A, D,$ and V) and gene-specific effects (G). Since G is global to $A, D,$ and V, gene-specific effects are orthogonal to global effects.

A simple ANOVA model for the factor main effect and the effects of interest, VG, is given as follows:

$$y_{ijkg} = \mu + A_i + D_j + V_k + G_g + (VG)_{kg} + \varepsilon_{ijkg}. \qquad (10.7.1)$$

The ANOVA model for the spot-to-spot variation by including AG effects is given here:

$$y_{ijkg} = \mu + A_i + D_j + V_k + G_g + (VG)_{kg} + (AG)_{ig} + \varepsilon_{ijkg}. \qquad (10.7.2)$$

The ANOVA model that includes the dye-gene interaction is given as follows:

$$y_{ijkg} = \mu + A_i + D_j + V_k + G_g + (VG)_{kg} + (AG)_{ig} + (DG)_{jg} + \varepsilon_{ijkg}. \qquad (10.7.3)$$

The additive error ε_{ijkg} is distributed as a normal distribution with mean 0 and variance σ^2.

We are interested in comparisons of varieties of fixed genes $(VG)_{1g} - (VG)_{2g}$.

In a design where variety k appears on r_k arrays, if we assume that the same set of genes is spotted on every array, VG interaction effects are orthogonal to all the other effects $A, D, V,$ and G. The least square estimate of $(VG)_{1g} - (VG)_{2g}$ could be shown to be

$$y_{..k_1 g} - y_{..k_2 g} - (y_{..k_1} - y_{..k}). \qquad (10.7.4)$$

For n genes, we could show that

$$\text{var}(y_{..k_1 g} - y_{..k_2 g} - (y_{..k_1} - y_{..k})) = \frac{n-1}{n}\left(\frac{1}{r_{k_1}} + \frac{1}{r_{k_2}}\sigma^2\right). \qquad (10.7.5)$$

We find that loop designs have a smaller efficiency if treatment numbers are large (Tempelman, 2005). If we consider a 12-treatment study based on a single classical loop such that $n = 2$ biological replicates are used per treatment, the resulting loop requires 12 arrays based on only 2 biological replicates per treatment. This design is presented as the loop 1 component where the treatments are connected in the following order:

$$A \rightarrow B \rightarrow C \rightarrow D \rightarrow E \rightarrow F \rightarrow G \rightarrow H \rightarrow I \rightarrow J \rightarrow K \rightarrow L \rightarrow A$$

The connected loop version using only Loop 1 would involve only a biological replicate per each of 12 treatments, and therefore, there will be no error degrees of freedom, which would make statistical inference impossible. Tempelman (2005) considered the classical loop ($n = 2$) design with a common reference design ($n = 2$) in which two biological replicates from each treatment are hybridized against a common reference sample for a total of 24 arrays.

In the interwoven loop design (Kerr and Churchill, 2001b) with two loops, we see that treatment arrangement in Loop 2 geometrically complements the treatment arrangement in Loop 1:

$$A \rightarrow F \rightarrow K \rightarrow D \rightarrow I \rightarrow B \rightarrow G \rightarrow L \rightarrow E \rightarrow J \rightarrow C \rightarrow H \rightarrow A$$

It is seen (Tempelman, 2005) that relative performance of nonreference designs generally exceeded that for reference design provided that the replication exceeded the minimally required $n = 2$. In all cases, the statistical efficiency is dependent on the number of biological replicates used and the residual variance ratio.

If an array is missed or damaged, it seems that reference design would be more flexible, but it is not the case always. If we consider n-array loop design (one array for each contrast) to the n-array reference design and one array fails in either design, in the loop design all the contrasts are still estimable, whereas in the reference design all the contrasts that involve the condition in the failed array are not estimable anymore. Moreover, multiple interwoven loops will make a loop design even more robust. If we have three conditions and each condition is measured four times, the loop design will have 12 slides. And if 25% of the 12 slides fail, any contrast can still be estimated, whereas in the case of the reference design, 25% of the total three slides will make the contrast unidentifiable. The effect of arrays on the balance of design is therefore the same in both loop and reference design.

## CHAPTER 11

Multiple Testing of Hypotheses

CONTENTS

11.1 INTRODUCTION

Genomic studies such as microarray, proteomic, and magnetic resonance imaging have provided a new class of large-scale simultaneous hypotheses testing problems in which thousands of cases are considered together. These testing problems require inference for high-dimensional multivariate distributions, with complex and unknown dependence structures among variables. These multivariate distributions have large parameter spaces that create problems in estimation. Moreover, the number of variables is much larger than the sample size. The multiple testing of hypotheses leads to an increase in false positives if correction procedures are not applied. In multiple hypothesis testing procedures, we test several null hypotheses with a single testing procedure and control the type I error rate. The parameters of interests may include regression coefficients in nonlinear models relating patient survival data to genome-wide transcript (i.e., mRNA) levels, DNA copy numbers, or SNP genotypes; measures of association between GO annotation and parameters of the distribution of microarray expression measures; or pairwise correlation coefficients between transcript levels (Dudoit, 2004a,b). In these problems, the number of null hypotheses varies from thousands to millions. These situations become more complicated when complex and unknown dependence structures among test statistics are present; e.g., Directed Acyclic Graph (DAG) structure of GO terms; Galois lattice for multilocus composite SNP genotypes.

11.2 TYPE I ERROR AND FDR

In multiple testing, the null hypothesis is rejected based on the cut-off rules for test statistics. This way, type error is fixed at a specified level. Since in most of the cases the distribution of the population is unknown, the null distribution is assumed to determine the cut-off values (*p*-values). It is done by generating a distribution from the given data under the null hypotheses (Westfall and Young, 1993).

One of the approaches for the multiple testing problems is to examine each case separately by a suitable procedure—for example, using a hypothesis testing procedure or confidence interval at a given level of significance. This approach is known as a case-by-case approach (in microarrays, it is called a gene-by-gene approach). Thus, for a paired two-sided test for $k \geq 3$ cases, one can perform $\binom{k}{2}$ pairwise two-sided tests, each at a preassigned significance level α. But such multiple tests do not account for the multiplicity selection effect (Tukey, 1977). If we apply $\binom{k}{2}$ individual tests for each hypothesis, the probability of finding any false positive will equal to α when exactly one false null hypothesis is true, and it will exceed α when two or more null hypotheses are true. Therefore, with the increase in k, the false positive increases. For example, we have 20,000 genes on a chip, and not a single gene is differentially expressed. The chance that at least one *p*-value is less than α when m independent tests are conducted is $1 - (1 - \alpha)^m$. If we set $\alpha = 0.01$ and $m = 10000$, then the chance is exactly 1, and if we set $\alpha = 0.01$ and $m = 1000$, then the chance is 0.9999568. Thus, a small p (unadjusted) value no longer leads to the right decision to reject the null hypothesis.

In hypothesis testing problems, errors committed can be classified as Type I errors or Type II errors. A Type I error, or false positive, is committed by rejecting a true null hypothesis, and a Type II error, or false negative, is committed by failing to reject a false null hypothesis. One would like to simultaneously minimize both types of errors, but due to the relation between two types of errors, this is not feasible. Therefore, it leads to fixing a more serious error in advance and minimization of the other. Typically, Type II errors are minimized; that is the maximization of power, subject to a Type I error constraint.

Now we consider the problem of testing simultaneously m null hypotheses $H_j, j = 1, \ldots, m$, and let R denote the number of rejected hypotheses. We form a table to summarize the four situations that occur in the testing of hypotheses (Benjamini and Hochberg, 1995). The specific m hypotheses are assumed to be known in advance; the numbers m_0 and $m_1 = m - m_0$ of true and false null hypotheses are unknown parameters; R is an observable random variable; and S, T, U, and V are unobservable random variables. In general, we would like to minimize the number V of false positives, or Type I errors, and the number T of false negatives, or Type II errors.

	# of hypotheses not rejected	# of hypotheses rejected	
# of true hypotheses	U	V	m_0
# of false hypotheses	T	S	m_1
	$m\text{-}R$	R	m

In a single hypothesis testing, the Type I error is generally controlled at some pre-assigned significance level α. We choose a critical value C_α for the test statistic T such that

$$P(|T| \geq C_\alpha | H_1) \leq \alpha, \text{ and we reject the null hypothesis if } |T| \geq \alpha. \quad (11.2.1)$$

Some of the most standard (Shaffer, 1995) error rates are defined as follows for the multiple comparison problem:

(a) Per-comparison error rate (PCER) is defined as

$$\text{PCER} = E(V)/m. \quad (11.2.2)$$

(b) Per-family error rate (PFER) is defined as

$$\text{PFER} = E(V). \quad (11.2.3)$$

(c) Family-wise error rate (FWER) is defined as the probability of at least one Type I error given as follows:

$$\text{FWER} = P(V \geq 1). \quad (11.2.4)$$

(d) False discovery rate (FDR; Benjamini and Hochberg, 1995) is defined as the expected proportion of Type I errors among the rejected hypotheses, i.e.,

$$\text{FDR} = E(Q), \quad (11.2.5)$$

where

$$Q = \begin{cases} \frac{V}{R}, & \text{if } R > 0, \\ 0, & \text{if } R = 0. \end{cases}$$

It is important to note that the expectations and probabilities given in the preceding equations are conditional on which hypotheses are true.

The Type I errors can be strongly controlled or weakly controlled. A strong control refers to a control of the Type I error rate under any combination of true and false hypotheses, whereas a weak control refers to a control of the Type I error rate only when all the null hypotheses are true. The FWER depends not only on which nulls are true and which are false in the hypothesis testing application, but also on the distributional characteristics of the data, including normality or lack thereof, and correlations among the test statistics.

In general, for a given multiple testing procedure (MTP),

$$\text{PCER} \leq \text{FWER} \leq \text{PFER.} \tag{11.2.6}$$

It is seen that the FDR depends on the joint distribution of the test statistics and, for a fixed procedure, FDR $\leq$ FWER, with FDR $=$ FWER under the complete null, since $V/R = 1$ when there is at least one rejection, and $V/R = 0$ when there are no rejections. Thus, in the case of overall null hypothesis, the expected value of V/R is just the probability of finding one significance. But in the case of a partial null configuration, FDR is always smaller than FWER.

Thus, if FWER $\leq \alpha$ for any MTP, then FDR $\leq \alpha$ for the same MTP. In most cases, FDR $\leq \alpha$ but still FWER $> \alpha$. Thus, in general, MTPs that control FDR are more powerful than MTPs that control FWER. Strong control of the FWER is required, such as Bonferroni procedures, which control the FWER in the strong sense, but the Benjamini and Hochberg (1995) procedure controls the FWER in the weak sense. The procedures that control PCER are generally less conservative than those procedures that control either the FDR or FWER, and tend to ignore the multiplicity problem altogether. The single hypothesis testing procedure can be extended to simultaneous testing of hypotheses by adjusting the p-value for each hypothesis. A multiple testing procedure may be defined in terms of adjusted p-values, denoted by $\tilde{p}$. Given any test procedure, the adjusted p-value corresponding to the test of a single hypothesis H_j can be defined as the level of the entire test procedure at which H_j would just be rejected, given the values of all test statistics (Hommel and Bernhard, 1999; Shaffer, 1995; Westfall and Young, 1993; Wright, 1992; Yekutieli and Benjamini, 1999).

The Bonferroni procedure rejects any H_j whose corresponding p-value is less than or equal to α/k.

For the Sidak method, the adjusted p-value is

$$\tilde{p}_j = 1 - (1 - p_j)^k. \tag{11.2.7}$$

For FWER, the FWER adjusted p-value for hypothesis H_j is

$$\tilde{p}_j = \inf \left\{ \alpha \in [0, 1] : H_j \text{ is rejected at FWER} = \alpha \right\}. \tag{11.2.8}$$

For the FDR (Yekutieli and Benjamini, 1999), the adjusted p-value is

$$\tilde{p}_j = \inf\{\alpha : H_j \text{ is rejected at FDR} = \alpha\}. \qquad (11.2.9)$$

Resampling methods have been used for some MTPs, which are described in terms of their adjusted p-values (Westfall and Young, 1993).

11.3 MULTIPLE TESTING PROCEDURES

There are basically three types of MTPs: single-step, step-down, and step-up procedures. In single-step procedures, each hypothesis is evaluated using a critical value that is independent of the results of tests of other hypotheses. In step-wise procedures, the rejection of a particular hypothesis is based not only on the total number of hypotheses, but also on the outcome of the tests of other hypotheses and hence results in the increase in power, while keeping the Type I error rate under control. In step-down procedures, the testing of hypotheses is done sequentially starting with the hypotheses corresponding to the most significant test statistics. One would stop testing the hypotheses once a hypothesis is accepted and all the remaining hypotheses are accepted automatically. In step-up procedures, the hypotheses correspond to the least significant test statistics tested successively, and once one hypothesis is rejected, all the remaining hypotheses are rejected automatically.

The single-step Bonferroni adjusted p-values are thus given by

$$\tilde{p}_j = \min\{mp_j, 1\}. \qquad (11.3.1)$$

The single-step Sidak adjusted p-values are given by

$$\tilde{p}_j = 1 - (1 - p_j)^m. \qquad (11.3.2)$$

If data are correlated, which is the case in genomics, the p-value computed will also be correlated and therefore need adjustments in MTP. Westfall and Young (1993) suggested an MTP that takes into account the dependent structure among test statistics based on adjusted p-value. The adjusted p-values are defined as

$$\tilde{p}_j = P\left(\min_{1 \leq k \leq m} P_k \leq p_j | H_0^C\right), \qquad (11.3.3)$$

where m is a variable for unadjusted p-value of the kth hypothesis and H_0^C is the complete null hypothesis.

As pointed out earlier, the single-step procedures are convenient to apply, but are conservative for the control of FWER. To improve the power but still retain the strong control of the FWER, one can use the step-down procedure. The Holm (1979) step-down procedure is less conservative than the Bonferroni method.

Holm's procedure is defined as follows.

Let $p_{(1)} \leq p_{(2)} \leq \ldots \leq p_{(m)}$ denote the observed ordered unadjusted p-values obtained for the corresponding hypotheses $H_{(1)}, H_{(2)}, \ldots, H_{(m)}$; then the thresholds are as follows:

$\frac{\alpha}{m}$ for the first gene, $\frac{\alpha}{m-1}$ for the second gene, and so on.

Let k be the largest of j, for which $p_{(j)} < \frac{\alpha}{m-j+1}$; then reject the null hypothesis $H_{(j)}, j = 1, 2, \ldots, m$. This can also be defined as

$$\tilde{p}_{(j)} = \max_{k=1,2,\ldots,j} \{\min((m-k+1)p_{(j)}, 1)\}. \tag{11.3.4}$$

This procedure ensures monotonicity of the adjusted p-value, which means that $\tilde{p}_{(1)} \leq \tilde{p}_{(2)} \leq \ldots \leq \tilde{p}_{(m)}$, and only one hypothesis can be rejected provided all hypotheses with smaller unadjusted p-values have already been rejected. The step-down Sidak adjusted p-values are defined as

$$\tilde{p}_{(j)} = \max_{k=1,2,\ldots,j} \left\{1 - (1 - p_{(j)})^{(m-k+1)}\right\}. \tag{11.3.5}$$

The Westfall and Young (1993) step-down adjusted p-value are defined by

$$\tilde{p}_{(j)} = \max_{k=1,2,\ldots,j} \left\{P\left(\min_{1\varepsilon\{(1),(2),\ldots,(m)\}} p_1 \leq p_{(k)}|H_0^C\right)\right\}. \tag{11.3.6}$$

The Holm's p-value can be obtained if, in Westfall and Young's (1993) step-down minP adjusted p-values, we make the assumption $P_1 \sim U[0, 1]$ and use the upper bound provided by Boole's inequality. Holm's procedure can be improved by using logically related hypotheses (Shaffer, 1986).

In the step-up procedure, the least significant p-value is identified, and most of the MTPs are based on Simes's (1986) probability results. The Simes inequality is defined as

$$P\left(p_{(j)} > \frac{\alpha j}{m}, \forall j = 1, \ldots\ldots, m|H_0^C\right) \geq 1 - \alpha. \tag{11.3.7}$$

The equality holds in continuous cases. Hochberg (1988) defined an MTP based on the Simes inequality, which was an improvement over Holm's procedure. Recall that for FWER at significance level α, we define

$$j^* = \max\left\{j : p_{(j)} \leq \frac{\alpha}{m-j+1}\right\}, \tag{11.3.8}$$

and reject hypotheses $H_{(j)}$, $j = 1, 2, \ldots, j^*$, and if no such j^* exists, no null hypothesis is rejected.

The step-up Hochberg adjusted p-values are given by

$$\tilde{p}_{(j)} = \min_{k=j,\ldots,m} \left\{ \min((m - k + 1)p_{(k)}, 1) \right\}. \tag{11.3.9}$$

The step-up procedures are generally more powerful than the corresponding step-down procedures, but all the MTPs based on Simes inequality assume the independence of test statistics and hence p-values. If these procedures are applied to dependent structures, then they will be more conservative MTPs and will lose power.

As pointed out earlier, Benjamini and Hochberg (1995) argued that the FWER procedures are more conservative and require tolerance of Type I errors if the Type I errors are less than the number of rejected hypotheses. Benjamini and Hochberg (1995) proposed a less conservative approach called FDR that controls the expected proportion of Type I errors among the rejected hypotheses. The Benjamini and Hochberg (1995) procedure for independent test statistics is defined as

$$j^* = \max\left\{ j : p_{(j)} \leq \frac{j\alpha}{m} \right\}, \tag{11.3.10}$$

reject $H_{(j)}$, $j = 1, 2, \ldots, j^*$. If no such j^* exists, no null hypothesis is rejected. The step-up adjusted p-value is given by

$$\tilde{p}_{(j)} = \min_{k=j,\ldots,m} \left\{ \min\left(\left(\frac{m}{k} p_{(k)}, 1 \right) \right) \right\}. \tag{11.3.11}$$

Benjamini and Yekutieli (2001) showed that the Benjamini and Hochberg (1995) procedure can be used to control FDR under certain dependent structures. However, FDR is less conservative than FWER, but a few false positives must be accepted as long as their number is small in comparison to the number of rejected hypotheses.

Tusher et al. (2001) considered Westfall and Young (1993) to be too stringent and proposed a method called Significance Analysis of Microarrays (SAM) for the microarray analysis which considered standard deviation of each gene expression for the relative change. In the case of SAM, which calculates FDR by permutation method, it is seen that for a small sample size, SAM gives a high false positive rate and may even break down for a small sample size. The second problem with SAM is the choice of number of permutations. The results vary greatly on the choice of number of permutations chosen. The third problem with the SAM procedure is that as the noise level increases, the false positive and false negative rates increase in the SAM procedure. Sha, Ye, and Pedro (2004) and Reiner et al. (2003) argued that in SAM, the estimated FDR is computed using permutations of the data, allowing the possibility of dependent tests; therefore, it is possible that this estimated FDR approximates the strongly controlled FDR when any subset of the null hypothesis is true. Since the number of distinct permutations is limited, the number of distinct p-values is limited. Thus, the FDR estimate turns out to be too "granular" so that either 0 or 300 significant genes are identified, depending on how the p-value was defined.

Lehmann and Romano (2005a) considered replacing control of FWER by controlling the probability of k or more false rejections, called k-FWER. They derived both single-step and step-down procedures that control k-FWER, without making any assumptions concerning the dependence structure of the p-values of the individual tests. The Lehmann and Romano (2005a) procedure is similar to the Hommel and Hoffman (1988) procedure. The problem with the k-FWER method is that one has to tolerate one or more false rejections, provided the number of false rejections is controlled. Romano and Wolf (2007) proposed the generalized version of the Lehmann and Romano (2005a) k-FWER procedure in finite and asymptotic situations using resampling methods. Efron (2007) suggested a simple empirical Bayes approach that allowed FDR analysis to proceed with a minimum of frequentist or Bayesian modeling assumptions. Efron's approach involves estimation of null density functions by using empirical methods. But the use of empirical null distribution reduces the accuracy of FDR, both in local and tail areas.

As noted previously, a common problem with multiple testing procedures designed to control the FWER is their lack of power when it involves large-scale testing problems, for example, in microarray setting. We need to consider some broader classes of Type I error rates that are less stringent than the FWER and may therefore be more appropriate for current high-dimensional applications. The gFWER is a relaxed version of the FWER, which allows $k > 0$ false positives; that is, gFWER(k) is defined as the probability of at least $(k + 1)$ Type I errors ($k = 0$ for the usual FWER). Dudoit et al. (2004) provided single-step procedures for control of the gFWER, and Van der Laan et al. (2004) provided an augmentation procedure for g-FWER, but not exact g-FWER. Korn et al. (2004) provided permutation-based procedures for the two-sample testing problem only. As mentioned earlier, Benjamini and co-authors proposed a variety of multiple testing procedures for controlling the false discovery rate (FDR), i.e., the expected value of the proportion of false positives among the rejected hypotheses that establish FDR control results under the assumption that the test statistics are either independently distributed or have certain forms of dependence, such as positive regression dependence (Benjamini and Yekutieli, 2001). In contrast to FDR-controlling approaches that focus on the expected value of the proportion of false positives among the rejected hypotheses, Genovese and Wasserman (2004) proposed procedures to control tail probabilities for this proportion, but under the assumption that the test statistics are independent, which were called confidence thresholds for the false discovery proportion (FDP).

■ Example 1

In an experiment, primary hepatocyte cells were exposed *in vitro* and rat liver tissue was exposed *in vivo* to understand how these systems compared in gene expression responses to RDX. Gene expressions were measured in primary cell cultures exposed to 7.5 mg/L RDX for 0 min, 7 min, 14 min, and 21 min in group A. Primary cell expression effects were compared to group B, of which the expressions in liver tissue were isolated from female Sprague-Dawley rats 0 min, 7 min, 14 min, and 21 min after gavage with 12 mg/Kg RDX. Samples were assessed within time point and cell type using 2-color, 8K gene microarrays. Table 11.3.1 gives the gene expression measurements.

Find adjusted *p*-values using Bonferroni's, Sidak's, and Holm's procedures.

Solution

The unadjusted *p*-values are obtained using two-sample independent t-tests assuming an equal variance approach. We assume that the genes are independent. We applied Bonferroni's, Sidak's, and Holm's procedures. The results are given in the Table 11.3.2.

Table 11.3.1 Gene Expression Measurements in Group A and Group B

Gene	Group A				Group B			
	0 min	7 min	14 min	21 min	0 min	7 min	14 min	21 min
1	30632.89	25832.6	22986.84	10604.9	38506.89	37536.13	34346.05	34383.23
2	1655.436	1399.905	852.4491	1196.051	4383.948	1804.219	2629.452	943.5066
3	1271.244	702.2323	677.083	632.021	594.9425	665.5299	938.4369	576.0553
4	1806.615	924.9635	792.9834	707.7102	776.0653	901.4204	763.8595	1269.593
5	4930.627	1383.655	2058.469	944.0863	1392.437	3325.969	3516.814	2631.809
6	1294.582	857.2069	951.6427	878.8396	817.1726	1269.056	1561.489	781.5454
7	16573.37	8858.784	9393.979	4711.884	13603.79	15811.49	15512.76	16007.95
8	2133.273	936.5962	1005.598	2393.399	1263.358	1934.957	7336.775	2342.045
9	1778.57	1380.669	984.5	1328.189	1029.803	1849.16	2168.396	1340.184
10	5593.917	6323.296	3186.013	2701.687	3698.456	4069.94	6203.822	4762.759
11	5545.008	5340.17	4365.895	3318.655	3899.78	4348.11	7417.764	4630.108
12	6783.493	7756.267	2885.957	1482.108	3288.319	6270.909	5164.337	5830.812
13	2569.262	2338.175	1539.732	1129.915	1783.889	2377.796	2776.452	2447.798
14	3172.84	1419.855	1325.588	1244.683	1185.362	1914.236	1680.782	3382.152
15	391.9491	332.4912	505.0763	763.5841	575.4447	443.9159	776.6825	413.4923
16	25218.1	6994.415	8187.154	3411.416	10893.81	22374.74	26847.4	23472.2
17	840.7887	697.8938	627.5033	844.4889	750.5089	758.0752	1155.18	533.146
18	2857.327	1277.064	2408.041	1075.572	3477.9	2835.033	1693.242	3467.418
19	1368.655	680.8053	643.2378	683.2998	652.6847	880.6184	649.9303	625.24
20	1161.22	632.9535	501.7743	507.1958	470.76	493.0973	394.6327	386.7777

Table 11.3.2 Adjusted p-Values

Gene	p-value Unadjusted	Adjusted p-values for Bonferroni	Adjusted p-value for Sidak Method	Ordered Unadjusted p-values	Holm's Adjusted p-value Method
1	0.02097	0.419407496	0.345489417	0.020970375	0.439938114
2	0.172993	1	0.977603145	0.0783585	1
3	0.489944	1	0.99999858	0.146701021	1
4	0.657617	1	1	0.146935842	1
5	0.716432	1	1	0.172992683	1
6	0.619406	1	0.999999996	0.296222829	1
7	0.078359	1	0.804457827	0.308670017	1
8	0.30867	1	0.999378118	0.327199941	1
9	0.477311	1	0.999997683	0.472351013	1
10	0.831506	1	1	0.477310852	1
11	0.664151	1	1	0.489943608	1
12	0.81086	1	1	0.619406303	1
13	0.296223	1	0.999111406	0.657616998	1
14	0.718302	1	1	0.664150848	1
15	0.683073	1	1	0.683072811	1
16	0.146936	1	0.958348188	0.716432262	1
17	0.751398	1	1	0.718301948	1
18	0.3272	1	0.999638823	0.751398362	1
19	0.472351	1	0.999997202	0.810859631	1
20	0.146701	1	0.958118279	0.831505785	1

Following is the R-code for example 1:

```
> #Example 1
> rm(list = ls())
> #install multest package
> # Enter p-values obtained from t-test
> rawp<-c(0.020970375, 0.172992683, 0.489943608,
0. 657616998,
+ 0.716432262, 0.619406303, 0.0783585, 0.308670017,
+ 0.477310852, 0.831505785, 0.664150848, 0.810859631,
+ 0.296222829, 0.718301948, 0.683072811, 0.146935842,
+ 0.751398362, 0.327199941, 0.472351013, 0.146701021)
>
> mt.rawp2adjp(rawp, proc = c("Bonferroni", "Holm",
"Hochberg", "SidakSS", "SidakSD", "BH", "BY"))
$adjp
rawp Bonferroni Holm Hochberg SidakSS SidakSD BH
[1,] 0.02097037 0.4194075 0.4194075 0.4194075 0.3454894
0.3454894 0.4194075
```

[2,] 0.07835850 1.0000000 1.0000000 0.8315058 0.8044578
0.7878327 0.6919707
[3,] 0.14670102 1.0000000 1.0000000 0.8315058 0.9581183
0.9424796 0.6919707
[4,] 0.14693584 1.0000000 1.0000000 0.8315058 0.9583482
0.9424796 0.6919707
[5,] 0.17299268 1.0000000 1.0000000 0.8315058 0.9776031
0.9521205 0.6919707
[6,] 0.29622283 1.0000000 1.0000000 0.8315058 0.9991114
0.9948533 0.8179999
[7,] 0.30867002 1.0000000 1.0000000 0.8315058 0.9993781
0.9948533 0.8179999
[8,] 0.32719994 1.0000000 1.0000000 0.8315058 0.9996388
0.9948533 0.8179999
[9,] 0.47235101 1.0000000 1.0000000 0.8315058 0.9999972
0.9995343 0.8315058
[10,] 0.47731085 1.0000000 1.0000000 0.8315058 0.9999977
0.9995343 0.8315058
[11,] 0.48994361 1.0000000 1.0000000 0.8315058 0.9999986
0.9995343 0.8315058
[12,] 0.61940630 1.0000000 1.0000000 0.8315058 1.0000000
0.9998324 0.8315058
[13,] 0.65761700 1.0000000 1.0000000 0.8315058 1.0000000
0.9998324 0.8315058
[14,] 0.66415085 1.0000000 1.0000000 0.8315058 1.0000000
0.9998324 0.8315058
[15,] 0.68307281 1.0000000 1.0000000 0.8315058 1.0000000
0.9998324 0.8315058
[16,] 0.71643226 1.0000000 1.0000000 0.8315058 1.0000000
0.9998324 0.8315058
[17,] 0.71830195 1.0000000 1.0000000 0.8315058 1.0000000
0.9998324 0.8315058
[18,] 0.75139836 1.0000000 1.0000000 0.8315058 1.0000000
0.9998324 0.8315058
[19,] 0.81085963 1.0000000 1.0000000 0.8315058 1.0000000
0.9998324 0.8315058
[20,] 0.83150579 1.0000000 1.0000000 0.8315058 1.0000000
0.9998324 0.8315058
BY
[1,] 1
[2,] 1
[3,] 1
[4,] 1
[5,] 1
[6,] 1

```
[7,] 1
[8,] 1
[9,] 1
[10,] 1
[11,] 1
[12,] 1
[13,] 1
[14,] 1
[15,] 1
[16,] 1
[17,] 1
[18,] 1
[19,] 1
[20,] 1
$index
[1] 1 7 20 16 2 13 8 18 19 9 3 6 4 11 15 5 14 17 12 10
>
```

Gene 1 is not statistically significant if we use only unadjusted p-values at $\alpha = 0.5$. In the case of multiple comparison, we adjust the p-value to get the correct value for the p-value. After applying the Bonferroni method, Sidak method, and Holm method, we find that gene 1 is not a statistically significant gene.

Thus, we find that it is extremely important to adjust the p-values or the α values when we have multiple comparisons; otherwise, it is highly possible to make a wrong inference about the expressed genes. The role of correlation structure of the genes is also important when we adjust the p-values for the multiple comparisons.

References

Alizadeh, A., Eisen, M. B., Davis, R. E., Ma, C., Lossos, I. S., Rosenwald, A., Boldrick, J. C., Sabet, H., Tran, T., Yu, X., Powell, J. I., Yang, L., Matri, G. E., Moore, T., Hudson, J., Lu, L., Lewish, D. B., Tibshrani, R., Sherlock, G., Chan, W. C., Greiner, T. C., Weise-burger, D. D., Armitrage, J. O., Warnke, R., Levy, R., Wilson, W., Grever, M. R., Byrd, J. C., Botstein, D., Brown, P. O., and Staudt, L. M. (2000). Distinct types of diffuse large B-cell lymphoma identified by gene expression profiling. *Nature*, **403**, 503–511.

Allison, D. B., Cui, X., Page, G. P., and Sabripour, M. (2006). Microarray data analysis: From dis-array to consolidation & consensus. *Nature Reviews Genetics*, **7**, 55–65.

Allison, D. B., Gadbury, G. L., Heo, M., Fernandez, J. R., Lee, C. K., Prolla, T. A., and Weindruck, R. (2002). A mixture model approach for the analysis of microarray gene expression data. *Computational Statistics and Data Analysis*, **39**, 1–20.

Altman, N. S., and Hua, J. (2006). Extending the loop design for two-channel microarray experiments. *Genetics Research*, **88**, 153–163.

Arfin, S. M., Long, A. D., Ito, E. T., Tolleri, L., Riehle, M. M., Paegle, E. S., and Hatfield, G. W. (2000). Global gene expression profiling in Escherichia coli K12. The effects of Integration host factor. *Journal of Biological Chemistry*, **275**, 29672–29684.

Arvestad, L., Berglund, A-C., Lagergren, J., and Sennblad, B. (2003). Bayesian gene/species tree reconciliation and orthology analysis using MCMC. *Bioinformatics*, **19**, Supplement 1, i7–i15.

Baldi, P., and Brunak, S. (2001). *Bioinformatics: The machine learning approach*, 2nd ed. Cambridge, MA: MIT Press.

Baldi, P., and Hatfield, G. W. (2002). *DNA microarrays and gene expression*, Cambridge, UK: Cambridge University Press.

Baldi, P., and Long, A. D. (2001). A Bayesian framework for the analysis of microarray expression data: Regularized t-test and statistical inferences of gene changes. *Bioinformatics*, **17**, 509–519.

Barash, Y., and Friedman, N. (2002). Context-specific Bayesian clustering for gene expression data. *Journal of Computational Biology*, **9**, 169–191.

Benjamini, Y., and Hochberg, Y. (1995). Controlling the false discovery rate: A practical and powerful approach to multiple testing. *Journal of the Royal Statistical Society, Series B*, **57**, 289–300.

Benjamini, Y., and Yekutieli, D. (2001). The control of the false discovery rate in multiple testing under dependency. *Annals of Statistics*, **29**, 1165–1188.

305

Bertone, P., and Snyder, M. (2005). Advances in functional protein microarrays. *FEBS Journal*, **272**, 5400–5411.

Bokka, S., and Mathur, S. K. (2006). A nonparametric likelihood ratio test to identify differentially expressed genes from microarray data. *Applied Bioinformatics*, **5**(4), 267–276.

Breitling, R., and Herzyk, P. (2005). Rank-based statistics as a nonparametric alternative of the t-statistic for the analysis of biological microarray data. *Journal of Bioinformatics and Computational Biology*, **3**, 1171–1189.

Brynildsen, M. P., Tran, L. M., and Liao, J. C. (2006). A Gibbs sampler for the identification of gene expression and network connectivity consistency, **22**(24), 3040–3046.

Butala, H. D., Ramakrishnan, A., and Sadana, A. (2003). A mathematical analysis using fractals for binding interactions of estrogen receptors to different ligands on biosensor surfaces. *Sensors and Actuators* B, **88**, 266–280.

Carter, T. A., Greenhall, J. A., Yoshida, S., Fuchs, S., Helton, R., Swaroop, A., Lockhart, D. J., and Barlow, C. (2005). Mechanisms of aging in senescence-accelerated mice. *Genome Biology*, 6:R48, doi:10. 1186/gb-2005–6–6–r48.

Casella, G., and George, E. I. (1992). Explaining the Gibbs sampler. *The American Statistician*, **46**(3), 167–174.

Chen, T., Filkov, V., and Skiena, S. (1999). Identifying gene regulatory networks from experimental data. *ACM-SIGAT. The Third Annual International Conference on Computational Molecular Biology* (RECOMB99) ACM press 1999; 93–103.

Chib, S., and Greenberg, E. (1995). Understanding the Metropolis-Hastings algorithm. *The American Statistician*, **49**, 327–335.

Churchill, G. A. (2002). Fundamentals of experimental design for cDNA microarrays. *Nature Genetics*, **32**, 490–495.

Crick, F. H. C. (1958). On protein synthesis. In *Symposium of the Society for Experimental Biology XII*, Robers, R. B. (Ed.) p.153. New York: Academic Press.

Cui, X., and Churchill, G. A. (2003). Statistical tests for differential expression in cDNA microarray experiments. *Genome Biology*, **4**, 210.

Davison, A. C., and Hinkley, D. V. (1997). *Bootstrap methods and their application.* Cambridge University Press: New York.

DeRisi, J. L., Iyer, V. R., and Brown, P. O. (1997). Exploring the metabolic and genetic control of gene expression on a genomic scale. *Science*, **278**, 680–685.

Dobbin, K., and Simon, R. (2002). Comparison of microarray designs for class comparison and class discovery. *Bioinformatics*, **8**, 1462–1469.

Dobbin, K., Shih, J. H., and Simon, R. (2003). Statistical design of reverse dye microarrays. *Bioinformatics*, **19**, 803–810.

Doke, A., Mathur, S. K., and Sadana, A. (2006). Fractal analysis of heart-related compounds. *Journal of Receptors and Signal Transduction*, **26**, 337–351.

Dudoit, S., Fridlyand, J., and Speed, T. P. (2002). Comparison of discrimination methods for the classification of tumors using gene expression data. *Journal of the American Statistical Association*, **97**, 77–87.

Dudoit, S., Shaffer, J. P., and Boldrick, J. C. (2003). Multiple hypothesis testing in microarray experiments. *Statistical Sciences*, **18**, 71–103.

Dudoit, S., Van Der Laan, M. J., and Birkner, M. D. (2004a). Multiple testing procedures for controlling tail probability error rates. *Working Paper 166*, Division of Biostatics, University of California, Berkeley.

Dudoit, S., Van Der Laan, M. J., and Pollard, K. S. (2004b). Multiple testing. I. Single step procedures for the controlling of general type I error rates. *Statistical Applications in Genetics and Molecular Biology*, **3**, article 13.

Dudoit, S., Yang, Y. H., Callow, M. J., and Speed, T. P. (2002). Statistical methods for identifying differentially expressed genes in replicated cDNA microarray experiments. *Statistica Sinica*, **12**(1) 111–139.

Efron, B. (2001). Robbins, empirical Bayes, and microarrays. *Technical Report No. 2001–30B/219*, Department of Statistics, Stanford University, Stanford, CA.

Efron, B. (2003). Robbins, empirical Bayes and microarrays. *Annals of Statistics*, **31**, 366–378.

Efron, B. (2006). Microarrays, empirical Bayes, and the two-groups model. http://www.stat.stanford.edu/~brad/papers/twogroups.pdf.

Efron, B. (2007). Correlation and large-scale simultaneous significance testing. *Journal of the American Statistical Association*, **102**, 93–103.

Efron, B. (2007). Size, power and false rates. *Annals of Statistics*, **35**(4), 1351–1377.

Efron, B., and Morris, C. (1973). Stein's estimation rule and its competitors—An empirical Bayes approach. *Journal of American Statistical Association*, **68**, 117–130.

Efron, B., and Morris, C. (1975). Data analysis using Stein's estimator and its generalizations. *Journal of American Statistical Association*, **70**, 311–319.

Efron, B., Storey, J. D., and Tibshirani, R. (2001). Microarrays, empirical Bayes methods, and false discovery rates. *Technical Report 2001–23B/217*, Department of Statistics, Stanford University, Stanford, CA.

Efron, B., Tibshirani, R., Goss, V., and Chu, G. (2000). Microarrays and their use in a comparative experiment. *Technical report*, Department of Statistics, Stanford University, Stanford, CA.

Efron, B., Tibshirani, R., Storey, J. D., and Tusher, V. (2001). Empirical Bayes analysis of a microarray experiment. *Journal of American Statistical Association*, **96**, 1151–1160.

Eisen, M. B., and Brown, P. O. (1999). DNA arrays for analysis of gene expression. *Methods in Enzymology*, **303**, 179–205.

Eisen, M., Spellman, P. T., Brown, P. O., and Botstein, D. (1998). Cluster analysis and display of genome-wide expression patterns. *Proceedings of the National Academy of Sciences, USA*, **95**, 14863–14868.

Ekins, R., and Chu, F. W. (1999). Microarrays: Their origins and applications. *Trends in Biotechnology*, **17**(6), 217–218.

Filkov, V., Skiena, S., and Zhi, J. (2002). Analysis techniques for microarray time-series data. *Journal of Computational Biology*, **9**(2), 317–330.

Frasor, J., Chang, E. C., Komm, B., Lin, C.-Y., Vega, V. B., Liu, E. T., Miller, L. D., Smeds, J., Bergh, J., and Katzenellenbogen, B. S. (2006). Gene expression preferentially regulated by tamoxifen in breast cancer cells and correlations with clinical outcome, *Cancer Research*, **66**, 7334–7340.

Friedman, N., Lineal, M., Nachman, I., and Pe'er, D. (2000). Using Bayesian networks to analyze expression data. *Journal of Computational Biology*, **7**, 601–620.

Gelfand, A. E., and Smith, A. F. M. (1990). Sampling-based approaches to calculating marginal densities. *Journal of the American Statistical Association*, **85**, 398–409.

Geman, S., and Geman, D. (1984). Stochastic relaxation, Gibbs distributions, and the Bayesian restoration of images. *IEEE Transactions on Pattern Analysis and Machine Intelligence*, **6**, 721–741.

Genovese, C. R., and Wasserman, L. (2004). A stochastic process approach to false discovery control. *Annals of Statistics*, **32**, 1035–1061.

Glonek, G. F., and Solomon, P. J. (2004). Factorial and time course designs for cDNA microarray experiments. *Biostatistics*, **5**, 89–111.

Golub, T. R., Slonim, D. K., Tamayo, P., Huard, C., Gaasenbeek, M., Mesirov, J. P., Coller, H., Loh, M., Downing, J. R., Caligiuri, M. A., Bloomeld, C. D., and Lander, E. S. (1999). Molecular classification of cancer: Class discovery and class prediction by gene expression monitoring. *Science*, **286**, 531–537.

Green, P. J. (1995). Reversible jump Markov chain Monte Carlo computation and Bayesian model determination. *Biometrika*, **82**, 711–732.

Guo, W., Sarkar, S., and Peddada, S. D. (2009). Controlling false discoveries in multidimensional directional decisions, with applications to gene expression data on ordered categories. *Biometrics*, in press.

Hall, D. A., Ptacek, J., and Snyder, M. (2007). Protein microarray technology. *Mechanisms of Ageing and Development*, **128**(1), 161–167.

Hall, D. A., Zhu, H., Royce, T., Gerstein, M., and Snyder, M. (2004). Regulation of gene expression by a metabolic enzyme. *Science*, **306**, 482–484.

Hastie, T., Tibshirani, R., and Friedman, J. (2001). *The elements of statistical learning*. New York: Springer.

Hastings, W. K. (1970). Monte Carlo sampling methods using Markov chains and their applications. *Biometrika*, **57**, 97–109.

Havlin, S. (1989). Molecular diffusion and reaction in the fractal approach to heterogeneous chemistry. In D. Avnir (ed.). *Surfaces, colloids, polymers.* (pp. 251–269). New York: John Wiley & Sons.

Hochberg, Y. (1988). A sharper Bonferroni procedure for multiple tests of significance. *Biometricka*, **75**, 800–802.

Hochberg, Y., and Tamhane, A. C. (1987). *Multiple comparisons procedures.* New York: Wiley.

Holm, S. (1979). A simple sequential rejective multiple test procedure. *Scandinavian Journal of Statistics*, **6**, 65–70.

Hommel, G. (1988). A stagewise rejective multiple test procedure based on a modified Bonferroni test. *Biometrika*, **75**, 383–386.

Hommel, G., and Bernhard, G. (1999). Bonferroni procedures for logically related hypotheses. *Journal of Statistical Planning and Inference*, **82**, 119–128.

Hommel, G., and Hoffman, T. (1988). Controlled uncertainty. In P. Bauer, G. Hommel, and E. Sonnemann (eds.). *Multiple hypothesis testing* (pp. 154–161). Heidelberg: Springer.

Hu, J. (2008). Cancer outlier detection based on likelihood ratio test. *Bioinformatics*, **24**(19), 2193–2199.

Hunter, L., Taylor, R. C., Leach, S. M., and Simon, R. (2001). GEST: A gene expression search tool based on a novel Bayesian similarity metric. *Bioinformatics*, **17**, S115–S122.

Husmeier, D., and McGuire, G. (2002). Detecting recombination with MCMC. *Bioinformatics*, **18**, S345–S353.

Hwang, J. T. G., and Peddada, S. D. (1994). Confidence interval estimation subject to order restrictions. *Annals of Statistics*, **22**, 67–93.

Ihaka, R., and Gentleman, R. (1996). R: A language for data analysis and graphics. *Journal of Computational and Graphical Statistics*, **5**, 299–314.

Imbeaud, S., Graudens, E., Boulanger, V., Barlet, X., Zaborski, P., Eveno, E., Mueller, O., Schroeder, A., and Auffray, C. (2005). Towards standardization of RNA quality assessment using user-independent classifiers of microcapillary electrophoresis traces. *Nucleic Acids Research*, **30**, e56.

Ishwaran, H., and Rao, J. S. (2003). Detecting differentially expressed genes in microarrays using Bayesian model selection. *Journal of the American Statistical Association, Theory and Methods*, **98**, 438–455.

Ishwaran, H., and Rao, J. S. (2005a). Spike and slab gene selection for multigroup microarray data. *Journal of the American Statistical Association*, **100**, 764–780.

Ishwaran, H., and Rao, J. S. (2005b). Spike and slab variable selection: Frequentist and Bayesian strategies. *Annals of Statistics*, **33**, 730–773.

Iyer, V. R., Eisen, M. B., Ross, D. T., Schuler, G., Moore, T., Lee, J. C. F., Trent, J. M., Staudt, L. M., Hudson, J. Jr., Boguski, M. S., Lashkari, D., Shalon, D., Botstein, D., and Brown, P. O. (1999). The transcriptional program in the response of human fibroblasts to serum. *Science*, **283**, 83–87.

Ji, L., and Tan, K. L. (2005). Identifying time-lagged gene clusters using gene expression data. *Bioinformatics*, **21**, 509–516.

Joshi, A., Van de Peer, Y., and Michoel, T. (2008). Analysis of a Gibbs sampler method for model-based clustering of gene expression data. *Bioinformatics*, **24**(2), 176–183.

Kariwa, H., Isegawa, Y., Arikawa, Takashima, I., Ueda, S., Yamanishi, K., and Hashimoto, N. (1994). Comparison of nucleotide sequences of M genome segments among Seoul virus strains isolated from Eastern Asia. *Virus Research*, **33**(1), 27–38.

Kepler, T. B., Crosby, L., and Morgan, K. T. (2002). Normalization and analysis of DNA microarray data by self-consistency and local regression. *Genome Biology*, **3**, RESEARCH0037.

Kerr, M. K., and Churchill, G. A. (2001a). Bootstrapping cluster analysis; assessing the reliability of conclusions from microarray experiments. *Proceedings of the National Academy of Sciences*, USA, **98**, 8961–8965.

Kerr, M. K., and Churchill, G. A. (2001b). Experimental design for gene expression microarrays. *Biostatistics*, **2**, 183–201.

Kerr, M. K., Martin, M., and Churchill, G. A. (2000). Analysis of variance in microarray data. *Journal of Computational Biology*, **7**, 819–837.

Kerr, M. K., Martin, M., and Churchill, G. A. (2001). Statistical design and the analysis of gene expression microarrays. *Genetical Research*, **77**, 123–128.

Kopelman, R. (1998). Fractal reaction kinetics, *Science*, **241**, 1620–1626.

Korn, E. L., Troendle, J. F., Mcshane, L. M., and Simon, R. (2004). Controlling the number of false discoveries: Application to high dimensional genomic data. *Journal of Statistical Planning Inference*, **124**, 379–398.

Kwon, A. T., Hoos, H. H., and Ng, R. (2003). Inference of transcriptional regulation relationships from gene expression data. *Bioinformatics*, **19**, 905–912.

Landgrebe, J., Bretz, F., and Brunner, E. (2004). Efficient two-sample designs for microarray experiments with biological replications. *In Silico Biology*, **4**, 461–470.

Lehmann, E. L., and Romano, J. P. (2005a). Generalizations of the familywise error rate. *Annals of Statistics*, **33**, 1138–1154.

Lehmann, E. L., and Romano, J. P. (2005b). *Testing statistical hypothesis*, 3rd ed. New York: Springer.

Lehmann, E. L., Romano, J. P., and Shaffer, J. P. (2005). On optimality of stepdown and stepup multiple test procedures. *Annals of Statistics*, **33**, 1084–1108.

Lin, K. K., Chudova, D., Hatfield, G. W., Smyth, P., and Andersen, B. (2004). Identification of hair cycle-associated genes from time-course gene expression profile data by using replicate variance. *Procedures of the National Academy of Sciences*, **101**, 15955–15960.

Liu, D., Weinberg, C. R., and Peddada, S. D. (2004). A geometric approach to determine association and coherence of the activation times of cell-cycling genes under differing experimental conditions. *Bioinformatics*, **20**, 2521–2528.

Long, A. D., Mangalam, H. J., Chan, B. Y. P., Tolleri, L., Hatfield, G. W., and Baldi, P. (2001). Gene expression profiling in Escherichia coli K12: Improved statistical inference from DNA microarray data using analysis of variance and a Bayesian statistical framework. *Journal of Biological Chemistry*, **276**, 19937–19944.

Maddox, B. (2002). *Rosalind Franklin: The Dark Lady of DNA*, HarperCollins, NY.

Mathur, S. K. (2005). A robust statistical method for identification of differentially expressed genes. *Applied Bioinformatics*, **4**(4), 247–252.

Mathur, S. K. (2009). A run based procedure to identify time-lagged gene clusters in microarray experiments. *Statistics in Medicine*, **28**(2), 326–337.

Mathur, S. K., Doke, A., and Sadana, A. (2006). Identification of hair cycle-associated genes from time-course gene expression profile using fractal analysis. *International Journal of Bioinformatics Research and Its Applications*, **2**(3), 249–258.

Mathur, S. K., and Dolo, S. (2008). A new efficient statistical test for detecting variability in the gene expression data. *Statistical Methods in Medical Research*, **17**, 405–419.

Metropolis, N., Rosenbluth, A. W., Rosenbluth, M. N., Teller, A. H., and Teller, E. (1953). Equations of state calculations by fast computing machines. *Journal of Chemical Physics*, **21**, 1087–1091.

Miklós, I. (2003). MCMC genome rearrangement. *Bioinformatics*, **19**(2), 130–137.

Miller, R. G., Jr. (1981). *Simultaneous statistical inference*, 2nd ed. New York: Springer.

Nadon, R., and Shoemaker, J. (2002). Statistical issues with microarrays: Processing and analysis. *Trends in Genetics*, **18**(5), 265–271.

National Institute of Neurological Disorders and Stroke (2008). Effect of atorvastatin (Lipitor) on gene expression in people with vascular disease. URL: http://clinicaltrials.gov/ct2/show/NCT00293748.

Newton, M. A., Kendziorski, C. M., Richmond, C. S., Blattner, F. R., and Tsui, K. W. (2001). On differential variability of expression ratios: Improving statistical inference about gene expression changes from microarray data. *Journal of Computational Biology*, **8**, 37–52.

Paik, S., Tang, G., Shak, S., Kim, C., Baker, J., Kim, W., Cronin, M., Baehner, F. L., Watson, D., Bryant, J., Costantino, J. P., Geyer, C. E., Jr., Wickerham, D. L., and Wolmark, N. (2006). Gene expression and benefit of chemotherapy in women with node-negative, estrogen receptor–positive breast cancer. *Journal of Clinical Oncology*, **24**(23), 3726–3734.

Pan, W., Lin, J., and Le, C. (2001). A mixture model approach to detecting differentially expressed genes with microarray data, *Report 2001–011*, Division of Biostatistics, University of Minnesota.

Park, T., Yi, S. G., Lee, S., Lee, S. Y., Yoo, D. H., Ahn, J. I., and Lee, Y. S. (2003). Statistical tests for identifying differentially expressed genes in time-course microarray experiments. *Bioinformatics*, **19**(6), 694–703.

Pearson, K. (1920). Notes on the History of Correlation, *Biometrika*, **13**(1), 25–45.

Peddada, S. D., Harris, S., Zajd, J., and Harvey, E. (2005). ORIOGEN: Order restricted inference for ordered gene expression data. *Bioinformatics*, **21**(20), 3933–3934.

Peddada, S. D., Lobenhofer, L., Li, L., Afshari, C., Weinberg, C., and Umbach, D. (2003). Gene selection and clustering for time-course and dose-response microarray experiments using order-restricted inference. *Bioinformatics*, **19**, 834–841.

Peddada, S. D., Prescott, K., and Conaway, M. (2001). Tests for order restrictions in binary data. *Biometrics*, **57**, 1219–1227.

Perkins, E. J., Bao, W., Guan, X., Ang, C. Y., Wolfinger, R. D., Chu, T. M., Meyer, S. M., and Inouye, L. S. (2006). Comparison of transcriptional responses in liver tissue and primary hepatocyte cell cultures after exposure to hexahydro-1, 3, 5–trinitro-1, 3, 5–triazine. *BMC Bioinformatics*, **S22**, doi: 10.1186/1471–2105–7–S4–S22.

Pollard, K. S., and Van Der Laan, M. J. (2003). Multiple testing for gene expression data: An investigation of null distributions with consequences for the permutation test. In *Proceedings of the 2003 International Multi-Conference in Computer Science and Engineering, METMBS'03 Conference 3–9*.

Pounds, S., and Cheng, C. (2004). Improving false discovery rate estimation. *Bioinformatics*, **20**, 1737–1745.

Pritchard, C. C., Hsu, L., Jeffrey, D., and Peters, N. (2001). Project normal: Defining normal variance in mouse gene expression. *PNAS*, **98**(23), 13266–13271.

R Project for Statistical Computing. At: http://www.r-project.org/.

Rao, C. R. (2002). *Linear statistical inference and its application*, 2nd ed. Wiley-Interscience. New York.

Raychaudhuri, S., Stuart, J. M., and Altman, R. B. (2000). Principal components analysis to summarize microarray experiments: Application to sporulation time series. *Pacific Symposium in Biocomputing 2000*, 455–466.

Reiner, A., Yekutieli, D., and Benjamini, Y. (2003). Identifying differentially expressed genes using false discovery rate controlling procedures. *Bioinformatics*, **19**(3), 368–375.

Robbins, H. (1951). Asymptotically sub-minimax solutions of compound statistical decision problems. *Proceedings of Second Berkeley Symposium*, **1**, 131–148. University of California Press.

Robbins, H. (1964). The empirical Bayes approach to statistical decision problems. *Annals of Mathematical Statistics*, **35**, 1–20.

Robbins, H., and Hannan, J. (1955). Asymptotic solution of the compound decision problem for two completely specified distributions. *Annals of Mathematical Statistics*, **26**, 37–51.

Robert, C. P., and Casella, G. (1998). *Monte Carlo statistical methods*. New York: Springer.

Roberts, G. O., Gelman, A., and Gilks, W. R. (1997). Weak convergence and optimal scaling of the random walk Metropolis algorithms. *Annals of Applied Probability*, **7**, 110–220.

Robertson, D. L., Hahn, B. H., and Sharp, P. M. (1995). Recombination in AIDS viruses. *Journal of Molecular Evolution*, **40**, 249–259.

Romano, J. P., and Shaikh, A. M. (2006). Stepup procedures for control of generalizations of the familywise error rate. *Annals of Statistics*, **34**, 1850–1873.

Romano, J. P., and Wolf, M. (2005a). Exact and appropriate step-down methods for multiple hypothesis testing. *Journal of the American Statistical Association*, **100**, 94–108.

Romano, J. P., and Wolf, M. (2005b). Stepwise multiple testing as formalized data snooping. *Econometrica*, **73**, 1237–1282.

Romano, J. P., and Wolf, M. (2007). Control of generalized error rates in multiple testing. *Annals of Statistics*, **35**(4), 1378–1408.

Rosenthal, J. S. (2007). AMCMC: An R interface for adaptive MCMC. *Computational Statistics & Data Analysis*, **51**(12). URL: http://www.probability.ca/amcmc/.

Sadana, A. (2003). A fractal analysis of protein to DNA binding kinetics using biosensors. *Biosensors Bioelectron*, **18**, 985–997.

Sayre, A. (2000). *Rosalind Franklin and DNA*. W.W. Nortron & Co. NY.

Sen, K., and Mathur, S. K. (1997). A bivariate signed rank test for two sample location problem. *Communications in Statistics, Theory and Methods*, **26**(12), 3031–3050.

Sen, K., and Mathur, S. K. (2000). A test for bivariate two sample location problem. *Communications in Statistics, Theory and Methods*, **29**(2), 417–436.

Sen, P. K. (2006). Robust statistical inference for high-dimensional data models with applications to genomics. *Austrian Journal of Statistics and Probability*, **35**(2&3), 197–214.

Sen, P. K., Tsai, M.-T., and Jou, Y. S. (2007). High-dimension low sample size perspectives in constrained statistical inference: The SARSCOVRNA genome illustration. *Journal of the American Statistical Association*, **102**, 685–694.

Sha, K. Ye, and Pedro, M. (2004). Evaluation of statistical methods for cDNA microarray differential expression analysis. *Eighth Annual International Conference on Research in Computational Molecular Biology (RECOMB 2004)*.

Shaffer, J. P. (1986). Modified sequentially rejective multiple test procedures. *Journal of the American Statistical Association*, **81**, 826–831.

Shaffer, J. P. (1995). Multiple hypothesis testing. *Annual Review of Psychology*, **46**, 561–584.

Sidwell, G. M., Jessup, G. L., Jr., and McFadden, W. D. (1956). Estimation of clean fleece weight from small side samples and from wool density, body weight, staple length and grease fleece weight. *Journal of Animal Science*, **15**, 218.

Simes, R. J. (1986). An improved Bonferroni procedure for multiple tests of significance. *Biometrika*, **73**, 751–754.

Simmons, S., and Peddada, S. D. (2007). Order-restricted inference for ordered gene expression (ORIOGEN) data under heteroscedastic variances. *Bioinformation*, **1**(10), 414–419.

Simon, R., and Dobbin, K. (2003). Experimental design of DNA microarray experiments. *Biotechniques*, March Supplement, 16–21.

Spirtes, P., Glymour, C., and Scheines, R. (2000). Constructing Bayesian Network models of gene expression networks from microarray data. *In Proceedings of the Atlantic Symposium on Computational Biology, Genome Information Systems & Technology 2000*.

Sreekumar, A., Nyati, M. K., Varambally, S., Barrette, T. R., Ghosh, D., Lawrence, T. S., and Chinnaiyan, A. M. (2001). Profiling of cancer cells using protein microarrays: Discovery of novel radiation-regulated proteins. *Cancer Research*, **61**, 7585–7593.

Tempelman, R. J. (2005). Assessing statistical precision, power, and robustness of alternative experimental designs for two color microarray platforms based on mixed effects models. *Veterinary Immunology and Immunopathology*, **105**, 175–186.

Tsodikov, A., Szabo, A., and Jones, D. (2002). Adjustments and measures of differential expression for microarray data. *Bioinformatics*, **18**, 251–260.

Tu, Y., Stolovitzky, G., and Klein, U. (2002). Quantitative noise analysis for gene expression microarray experiments. *Proceedings of the National Academy of Sciences*, **99**, 14031–14036.

Tukey, J. W. (1977). *Exploratory data analysis*. Reading, MA: Addison-Wesley.

Tusher, V. G., Tibshirani, R., and Chu, G. (2001). Significance analysis of microarrays applied to transcriptional responses to ionizing radiation. *Proceedings of the National Academy of Sciences*, **98**, 5116–5121.

U. S. Department of Energy and the National Institutes of Health. The Human Genome Project (HGP). URL: http://www.ornl.gov/sci/techresources/Human_Genome/home.shtml.

Van Der Laan, M. J., Birkner, M. D., and Hubbard, A. E. (2005). Empirical Bayes and resampling based multiple testing procedure controlling tail probability of the proportion of false positives. *Statistical Applications in Genetics and Molecular Biology*, **4**, article 29.

Van Der Laan, M. J., Dudoit, S., and Pollard, K. S. (2004). Augmentation procedures for control of the generalized family-wise error rate and tail probabilities for the proportion of false positives. *Statistics in Applied Genetics and Molecular Biology*, **3**, article 15.

van't Veer, L. J., Dai, H., van de Vijver, M. J., He, Y. D., Hart, A. A., Mao, M., Peterse, H. L., van der Kooy, K., Marton, M. J., Witteveen, A. T., Schreiber, G. J., Kerkhoven, R. M., Roberts, C., Linsley, P. S., Bernards, R., and Friend, S. H. (2002). Gene expression profiling predicts clinical outcome of breast cancer. *Nature*, **415**(6871), 530–536.

Vinciotti, V., Khanin, R., D'Alimonte, D., Liu, X., Cattini, N., Hotchkiss, G., Bucca, G., de Jesus, O., Rasaiyaah, J., and Smith, C. P. (2005). An experimental evaluation of a loop versus a reference design for two-channel microarrays. *Bioinformatics*, **21**, 492–501.

Waagepetersen, R., and Sorensen, D. (2001). A tutorial on reversible jump with a view toward applications in QTL-mapping. *International Statistical Review*, **69**, 49–61.

Wang, Y., Lu, J., Lee, R., Gu, Z., and Clarke, R. (2002). Iterative normalization of cDNA microarray data. *IEEE Transactions on Information Technology in Biomedicine*, **6**, 29–37.

Wang, S., and Ethier, S. (2004). A generalized likelihood ratio test to identify differentially expressed genes from microarray data. *Bioinformatics*, **20**(1), 100–104.

Watson, J. D., and Crick, F. H. C. (1953). A structure for DNA. *Nature*, **171**, 737–738.

Welsh, J. B., Zarrinkar, P. P., Sapinoso, L. M., Kern, S. G., Behling, C. A., Monk, B. J., Lockhart, D. J., Burger, R. A., and Hampton, G. M. (2001). Analysis of gene expression profiles in normal and neoplastic ovarian tissue samples identifies candidate molecular markers of epithelial ovarian cancer. *Proceedings of the National Academy of Sciences*, **98**, 1176–1181.

Wernisch, L. (2002). Can replication save noisy microarray data? *Comparative and Functional Genomics*, **3**, 372–374.

Wernisch, L., Kendall, S. L., Soneji, S., Wietzorrek, A., Parish, T., Hinds, J., Butcher, P. D., and Stoker, N. G. (2003). Analysis of whole-genome microarray replicates using mixed models. *Bioinformatics*, **19**, 53–61.

Westfall, P. H., and Young, S. S. (1993). *Resampling-based multiple testing: Examples and methods for p-value adjustment*. New York: Wiley.

Winer, B. J. (1971). *Statistical principles in experimental design*, 2nd ed., New York: McGraw-Hill.

Wright, S. P. (1992). Adjusted *p*-values for simultaneous inference. *Biometrics*, **48**(4), 1005–1013.

Xiong, M., Li, W., Zhao, J., Jin, L., and Boerwinkle, E. (2001). Gene selection in gene expression based tumor classification. *Molecular Metabolism*, **73**, 239–247.

Yang, Y. H., Dudoit, S., Luu, P., and Speed, T. P. (2001). Normalization for cDNA microarray data. In M. L. Bittner, Y. Chen, A. N. Dorsel, and E. R. Dougherty (eds.), *Microarrays: Optical technologies and informatics*, **4266**, *Proceedings of SPIE*, San Diego, CA, 141–152.

Yang, Y. H., Buckley, M. J., Dudoit, S., and Speed, T. P. (2002). Comparison of methods for image analysis on cDNA microarray data. *Journal of Computational and Graphical Statistics*, **11**(1) 108–136.

Yang, Y. H., Dudoit, S., Luu, P., Lin, D. M., Peng, V., Ngai, J., and Speed, T. P. (2002). Normalization for cDNA microarray data: A robust composite method addressing single and multiple slide systematic variation. *Nucleic Acids Research*, **30**(4) e15.

Yates, F. (1935). *The design and analysis of factorial experiments*. Technical Communication no. 35 of the Commonwealth Bureau of Soils.

Yekutieli, D., and Benjamini, Y. (1999). Resampling-based false discovery rate controlling multiple test procedures for correlated test statistics. *Journal of Statistical Planning and Inference*, **82**, 171–196.

Index

315

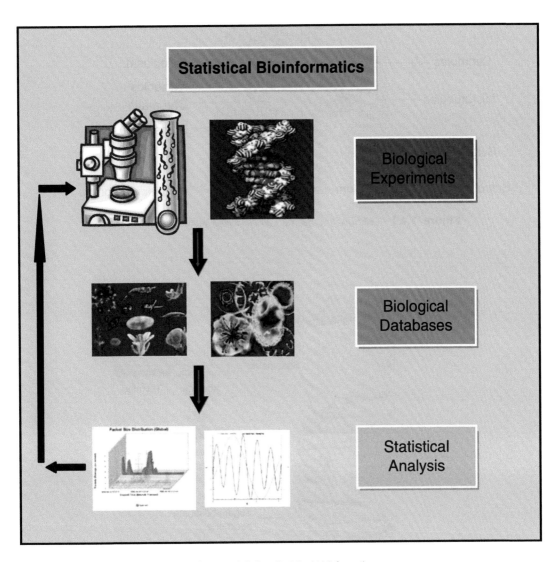

Figure 1.1.1 *Statistical bioinformatics.*

Cell Structure

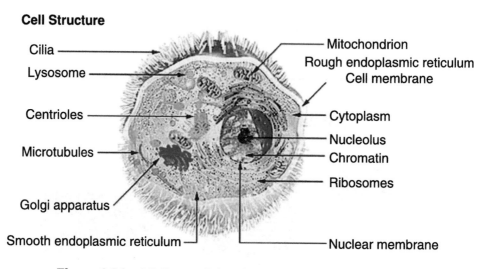

Cilia

Lysosome

Centrioles

Microtubules

Golgi apparatus

Smooth endoplasmic reticulum

Mitochondrion

Rough endoplasmic reticulum

Cell membrane

Cytoplasm

Nucleolus

Chromatin

Ribosomes

Nuclear membrane

Figure 1.4.1   *Cells (Courtesy: National Cancer Institute-National Institutes of Health).*

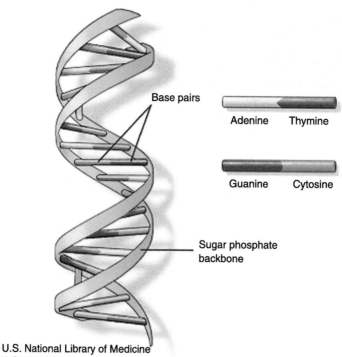

Base pairs

Adenine   Thymine

Guanine   Cytosine

Sugar phosphate
backbone

U.S. National Library of Medicine

Figure 1.5.2   *DNA.*

Printed and bound by CPI Group (UK) Ltd, Croydon, CR0 4YY

03/10/2024

01040316-0006

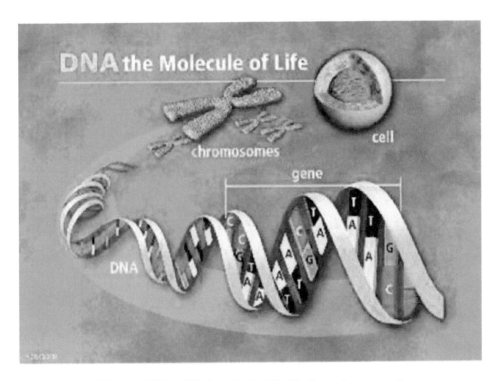

Figure 1.5.3 *DNA, the molecule of life (http://genomics.energy.gov).*

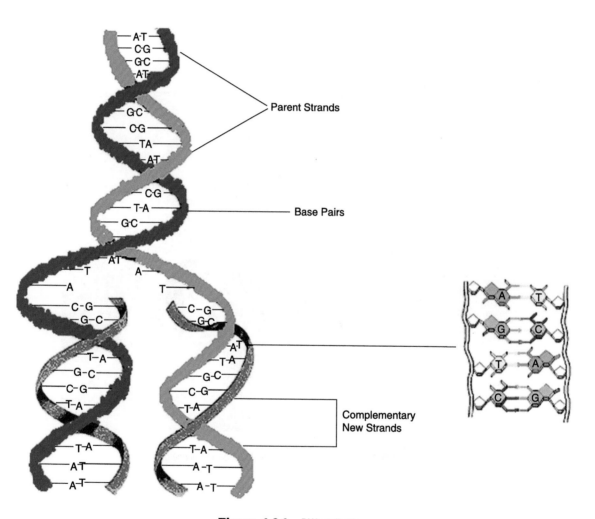

Figure 1.6.1 *DNA replication.*

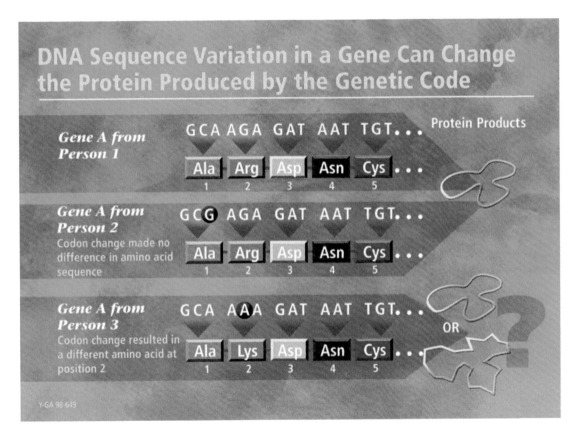

Figure 1.8.1 *DNA sequence (http://genomics.energy.gov).*

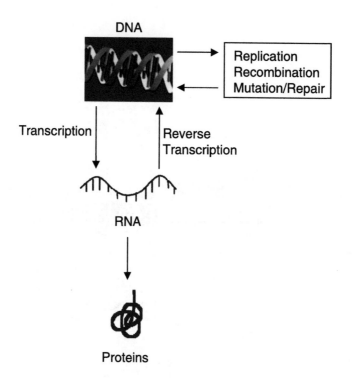

Figure 1.9.1 *Genetic flow of information.*

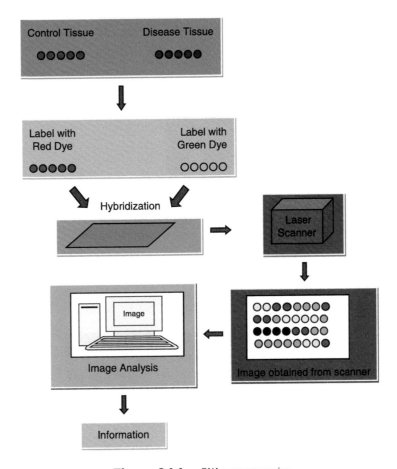

Figure 2.1.1 *cDNA array processing.*

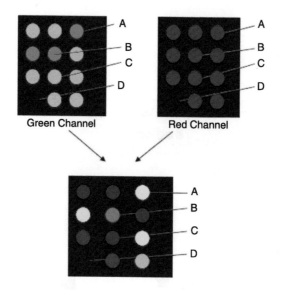

Figure 2.1.2 *Microarray slides.*

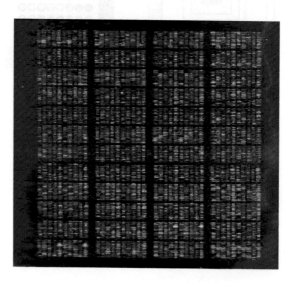

Figure 2.1.3 *Scanned image obtained from affymetric machine.*